MEETING MY MAKER

The Universal Mind and Reincarnation

BY

MICHAEL A. TEWELL, M.A., J.D.

PUBLISHED 2022

MEETING MY MAKER: The Universal Mind and Reincarnation.

Contact information:

Author: Michael A. Tewell, P.O. Box 1314, Palm Harbor, Florida, USA 34682

and the Universal Mind Foundation, Inc.

(a not-for-profit corporation incorporated under Florida law)

Website: universalmindfoundation.org. (Beginning May, 2022)

Email: universalmindtewell@gmail.com

Twitter: @UniversalMindF1

ISBN 978-0-9600607-8-8 (ebook)

ISBN 978-0-9600607-9-5 (paperback)

1. Universal Mind. 2. Spirituality. 3. Self Help.

4. New Age. 5. Science. 6. Personal Development.

PUBLISHED IN THE UNITED STATES OF AMERICA

BY MICHAEL A. TEWELL, ESQ.

TABLE OF CONTENTS

ACKNOWLEDGEMENTS

I am very grateful to my wife and son for their cherished love and support, without which, this work would have been impossible. They let me be me.

This book is also dedicated to you and all who search for true meaning in life. After all, it is for you that this book is written. Your opinion of this book is important.

Be willing to share your thoughts and reactions to what you read here with others on social media, book reviews and on the Universal Mind Foundation, Inc. website. Truth matters, now more than ever. In the process, you enhance everyone's awareness of the ultimate nature of both truth and consciousness. For consciousness, like truth, are realities we all share.

My social contact addresses are…

Michael A. Tewell, Esq., P.O. Box 1314, Palm Harbor, Florida, USA 34682
Website: universalmindfoundation.org.
Email: universalmindtewell@gmail.com
Twitter: @UniversalMindF1

INTRODUCTION

No one wants to die. Death is the one visitor in our lives which never requires an invitation and always arrives too soon. For the young, it tragically ends a life before it really begins. There is no greater tragedy than the death of a child. For the old, death ends relationships and accomplishments that took decades to achieve. For those in between, death unceremoniously interrupts a life full of accomplishment and unrealized potential and tears apart family and loved ones.

For the survivors, death of a loved one creates a bitter void that can never be resolved. This is the bitter reality of death. Christianity offers a salve to the pain of bereavement, but only to its believers. To believe in eternal bliss in a Christian heaven, one must accept its many myths, fear its false hell and kneel to its human authority.

Christianity is not the only sinner. Too many man-made religions prefer to prey upon our fear of death than seek to change the world for the better. Any religion which would rather keep its gold and silver while its people suffer from poverty, disease and human rights violations is NOT a religion worth following.

There is an alternative, however. This book presents the scientific-based theory that all life in the universe is the creation of a Universal Mind, and that Reincarnation is a causal consequence of universal laws.

This theory is NOT mere conjecture or speculation. This theory is based upon known and accepted scientific discoveries and research anyone can access. The source for each scientific discovery or research referenced in this book is documented by source citation so you, the reader, can be assured of its authenticity—or you can check the source yourself.

Each scientific discovery cited is like a piece of a puzzle. Separately, they do not seem relevant to each other. But when we put each piece together in its proper place, the individual puzzle pieces create a compelling picture of how we exist in the universe and how that existence takes us beyond our human-made religions into a new era of self-discovery and human evolution.

How does a retired criminal defense trial lawyer come to write books teaching people about the Universal Mind and Reincarnation? While still a young child, I had a Near-Death-Experience. I "died." Then I came back from the dead. What I experienced during this extraordinary event changed my life—and my perception of the universe.

I realize most readers of this book have little or no belief in a Universal Mind or Reincarnation. Their perception of the universe, and themselves, is the result of a lifetime of indoctrination by religious dogma created by men obsessed with Earthly-power and wealth.

All I ask is that you open your mind to the possibility you may be wrong. Do not be afraid to consider an alternative origin for what and who you are. If you are so certain your religious dogma is correct, what is there to fear by reading this book?

Fear is a mind killer. The only salvation from fear is knowledge of the truth. I challenge you to read this book and then decide which makes more sense—Christian myth or the science-backed theory presented in this book.

This book is also written for those whose life experience cannot be explained by religion. You are searching for answers which cannot be found in your Bible, Torah, Qur'an or other religious texts.

Do not consider me special—or that I have done something worthy of this book. I am not. I am but a flawed, imperfect human who has made his share of poor choices—as my ex-wives will be more than eager to point out. This book simply shares with you a few of my personal experiences which inspired me to search for answers beyond my religious teaching. It is my sincere hope that this book furthers your quest for truth as you seek your own purpose and meaning in this universe. This is the sole purpose of this book.

In my two previous books, ***REINCARNATION: The Scientific Theory*** and ***IMMORTALITY: The Science of Forbidden Fruit***, the reader is taken on a step-by-step journey of how evolution has adapted specific human physiological processes within the human brain to allow our individuality to be passed on from one live to the next—Reincarnation.

Now in ***MEETING MY MAKER: The Universal Mind and Reincarnation***, we step beyond the human physiological processes of Reincarnation to learn how and why Reincarnation is a natural consequence of the universe and its natural laws. We explore how the birth of our universe planted the seeds for our individuality; how Reincarnation is a causal consequence of universal laws; and the incredible purpose the universe has hidden within our minds.

I challenge you to open your mind to these possibilities as you read this book. Once you open yourself to the Universal Mind, your existence will be changed forever.

CHAPTER 1

When a Child Dies…

Ever wondered why you have those strange "waking dreams" just before you wake up in the morning? Those weird dreams that are too vivid, too graphic, too "real" to be just ordinary dreams? They are the dreams that take you back in time to places and events that could never have happened to you in this life; yet are still strangely familiar to you, like déjà vu.

Ever had dreams like these? If you have, your past-life memories are surfacing in your consciousness. Past-life memories are very difficult to reach. They are transplanted by photon-wave from your past body into the spinal tissue of your present body as neural signals during the fetal stage of your next life.

Once you are born, these neural-signals travel up the spinal cord to the brain where they are stored like any other long-term memory. They are then "covered over" by the billions of memories your brain stores over the years. Access these "deep-core" memories is not easy.

But there are three possible ways these past-life memories surface in our consciousness: (1) meditation can "still" the conscious mind from sensory stimuli and superficial, daily thoughts to open a connection to the buried, long-term storage areas of the brain holding the past-life memory neurons; (2) spontaneous moments of *déjà vu* brought on by some trigger such as a random encounter, emotionally traumatic event, a "Near Death

Experience"; or (3) a brain injury, such as a concussion or tumor which involves the region of the brain storing these specific past-life neurons.

This is often how memories of past lives reawaken within the deep recesses of the brain and begin to play out for us like scenes from a motion picture just before we wake up. We see everything, feel everything, just as if we were there in real time—only these dreams occur right before we become fully awake.

My own "waking dreams" and visions began when I was just a child of six-years-old. Looking back, I have to say that I had a unique opportunity to reach these "deep core" memories, despite my young age. There were two reasons for my being able to reach those past-life memories.

I was born in 1950 with chronic, severe asthma. This was a time when medical experts still did not fully understand what asthma was, what caused it, or how to mitigate its symptoms. Back then, too many asthma attacks ended in death. Severe asthma attacks often required epinephrine injections directly into the heart to save the patient.

Most take breathing for granted. Not so for anyone with severe asthma. Anyone who has suffered severe asthma knows what it feels like—an elephant sitting on your chest every minute of every day and every minute of every night.

I was so ill as a youngster that breathing required a conscious act of will—24/7—from toddlerhood to preschool. My mom kept me in a sheet-covered crib with steamed air to help me breathe. I must say this make-shift "tent" filled with steamed-air didn't really help.

Back then, medical professionals believed that asthma was due to a lack of oxygen into the lungs. This was a major medical error. Asthma is caused by the body's inability to process the oxygen from the lungs into the blood stream. Asthma is not caused by a lack of oxygen supply but a blockage within the oxygen-blood exchange cycle.

Sadly, the tent-steam gimmick only made me almost suffocate. But mom tried. What did work for me, I discovered purely by accident. It was my own anecdotal discovery, and it may not work for anyone else. Carbonated colas and soda crackers. The cola must be caramel in color. Clear sodas did not work for me. But, if they help you, fantastic.

As I grew from toddler to preschooler; stuck in that crib all-day, every day; I developed my own ways of taking my mind off the unremitting struggle to breathe. To do so, I would let my mind go blank for as long as I could.

Notice I said "let" and not "force." The use of force when attempting meditation or any other mental technique is counterproductive. The best way to relate to your deeper consciousness is to accept that it possesses a will of its own. Let it teach you.

Without knowing it at the time, this "mind-blanking" I put myself through to keep breathing during asthma attacks was a form of focused-meditation. This self-taught meditation must have opened a channel to my past-life memories despite my young age. But that probably wasn't the only reason.

By the time I was five-years-old, the doctors treating me in Ohio gave trying to make me better. Their "solution" was that my parents should move me out to Phoenix, Arizona where the climate was warmer and less humid. The doctors warned that if I stayed in Ohio, the Midwest winters would eventually kill me.

The Ohio physicians treating me had no idea they were sending me from a freezer into a blast furnace. Which is worse? Ohio's cold air freezing the bronchial tubes or the daily sandstorms that immerse the bronchial tubes in grains of dust so you cannot breathe at all?

Unaware of what faced us, my parents sold everything and moved me to Phoenix where we lived on Pinchot Avenue, next to Thomas Road. Our house was on the last street before the desert. We lived on the very edge of town. Past the back-yard fence, was nothing but desert and Camelback Mountain.

I was fine for a few weeks. I was making friends and having fun in first grade at Lafayette Elementary. But then, the sandstorms began to hit us almost daily. My health grew steadily worse. Then, I came down with double-pneumonia.

It was then that I had my "Near-Death Experience." Basically, "I died" from a severe asthma attack brought on by bout of severe pneumonia. Unaware of how ill I was, my parents made me go to school. My parents didn't want me to fail. I had just turned six.

Then, one day, I spent the time in class struggling to breathe. Fortunately, a concerned classmate and friend of my saw my face turning blue from loss of oxygen and told the teacher. Shocked, the teacher sent me home. I walked home only to pass out from oxygen starvation on my front porch.

My parents rushed me to the hospital. Grandpa drove with grandma in front. I was laid out on my mom's lap in the backseat. It was a bizarre trip. My mom cried. My grandpa kept telling her to shut up, he couldn't drive with her crying all the time.

I spent the time gazing out the car window. There was no energy for anything else. All my energy had been spent at school struggling to breathe. The thought occurred to me that there was no longer any need to struggle. I could just stop breathing and everything would just take care of itself. As my grandpa raced to the hospital, I just let my mind go and watched the street lights and telephone poles flash by.

Suddenly, I found myself staring at a female face. She was staring right back at me. She wasn't a ghost; she had human features: a head and long hair. Her facial features were that of a mid-30s woman with a kind, compassionate smile.

But there was no color. The face, the hair, the head were all white. Not the "white" of the Caucasian race, but white as in an acrylic painting using Titanium white paint, then washing it out until it is almost transparent.

She was beautiful, in a Classically Greek kind of way. I instantly adored her. The way she effortlessly floated along-side our car was mesmerizing. It was immediately obvious to me that this woman was no ordinary human.

But neither was she an angel like what we are taught in Sunday School. She had no wings. She mentioned no religious dogma or myth. There was no mention of heaven or hell, Jesus or Yeshua, Christianity, Judaism or Islam. If anything, she was totally non-religious.

She introduced herself by name, but I cannot recall what it was. I tried to keep up be after the third syllable I gave up trying to remember it. I was too sick and her name was too long for me to remember. I do recall that it started with an "Aera" sound.

"You don't have to breath anymore," she offered with a compassionate smile.

"You can come with me. We are sorry you were given such an unhealthy body. We will make sure you have a heathier body next life," she promised. "All you have to do is tell me you want to come with me."

I didn't know what to do. I was laying in my mother's lap in the back seat, looking out the window at this pale-white face staring at me as it floated along, outside the car, as my grandfather raced me to the hospital.

"Mom, there's an angel outside the car. She wants me to go with her."

Chaos ensued. Mom stopped crying and began screaming hysterically. Grandfather yelled at her to shut up. Mom screams gradually were replaced by occasional, choked sobs.

By the time we arrived at the hospital emergency entrance, I was floating outside my body. It occurred to me that this must be what how the ghostly-appearing woman floated just outside our car.

Being outside my physical body had a noticeable effect. Though my body was dying, my mind was sharp—perhaps even sharper than it had been before I stopped breathing. I was suddenly able to be aware of things I wouldn't have noticed before as an ordinary six-year-old.

I could see my parents in the waiting room standing alone. Grandpa was staring blankly at the ER double-doors. Grandma was staring at a wall. Mom was still sobbing. Each was alone in their thoughts. Their pain and despair was palpable.

The emergency room physician was a young man and was visibly taken aback by my condition. He was a young doctor, and I somehow "knew" he had come to work today expecting to catch up on his reading. When he came to work that morning, he had no clue he was going to have to fight for my life. I felt his uncertainly turning to panic.

"How long has he not been breathing?" He asked, as a nurse quickly took my body from my mother's arms and rushed it to the trauma table.

"Right when we came in," answered Grandpa. He was wrong, of course. I had stopped breathing before we arrived, but my mom was too upset to correct him.

My mind was no longer connected to my body lying on the trauma table. My consciousness was floating above the emergency room trauma-bed, looking down at the panicked young doctor trying to figure out why I was turning blue and not breathing; while my mom and grandparents stood helplessly, outside in the waiting area.

The young doctor fascinated me. He had a huge challenge on his hands. A medical diagnosis is like a puzzle and I was curious to find out whether he would figure it all out in time. His patient was blue and not breathing. Could he figure out what to do to get the patient breathing again before the patient died. I felt no emotion. My interest was entirely and objectively academic, floating just above the trauma lamps, watching the doctor stress out below me.

Then, without warning, a hole began growing in the ceiling above me. It was round, white in the center and its edges active like lightning. I thought, how odd, a hole in the ceiling. How are they going to save me with a hole in the ceiling?

From out of the center of the hole or portal, a procession of white orbs emerged. Ten to twelve in number, these orbs remained up by the ceiling. Then two more orbs emerged. They came down to where I was floating, just above the trauma lamps.

"We are sorry you have been given such an unhealthy body," they thought to me.

I knew it was telepathy. Logically, I knew I couldn't speak. My body was unconscious on the surgical table below us. There is no other possibility. Our communication was telepathic.

"You must decide now whether you wish to come with us," they urged. "Time is short. If you come with us, we will see that you get a much healthier body next life. Do you want to go or stay?"

They didn't spell it out for me and they didn't need to. Though only 6-years-old, I knew somehow what was meant by "go" and "stay." What also got my attention was their telling me I had to decide immediately. If I didn't decide, they were going to take me away anyway. They did not explain why, but time was an issue for them (I didn't know then the relationship between loss of breathing and brain function).

I didn't really want to stay. I was sick and tired of being sick and tired all the time. Fighting to breathe all day gets old. Most asthma attacks are of the acute form. They hit you, but you get over it. The problem for me was that I had severe chronic asthma—the hell never ended. I wouldn't wish one of my asthma attacks on my worst enemy. Once you've been through torture, you don't want any else to suffer—not even an enemy.

But I chose to stay. As I was floating around, I got a chance to watch my family as they stood in the ER waiting room. They were suffering in silence; already grieving.

I felt sorry for them. I knew that if I should die, they would forever blame themselves. My dying so young would tear them apart as a family. I did not want that to be my last act in this life.

I am not a quitter. Asthma turned a sickly, little kid into a fearless and stubborn fighter. Life is precious. Every life is precious and not to be wasted or abused. Given the opportunity to stay, I took it.

Within seconds, the ER doctor plunged a hypo deep into my chest. To my utter amazement I didn't feel it. Instantly, I'm lying on the surgical table, back into my body. It was incredible! I was breathing again! Exhausted, but alive!

I spent the next two weeks at home, away from school. The doctors said my body needed time to recuperate. They said that "dying" and coming back might change my body chemistry somehow and they weren't sure how to handle it.

What the doctors did not know—what they could not know—was that it was not only my body that had been altered. As my body worked to heal itself, so too did my consciousness. In fact, my consciousness was affected at least as much as was my body—probably more so.

During those first few weeks after my NDE, there were significant changes in my behavior and sense of self. My consciousness was "off." I felt out of phase within myself, as though my mind were in two places at once.

I experienced a significant sense of detachment from this life, yet a feeling of oneness with everything. I felt so distant to my surroundings, my family, yet so in tune with an over-arching plane of existence. I didn't speak much. I kept to myself. Late at night, I would wake up, sneak outside just to look up at the night-time sky and stare up at the stars. It was very strange.

Concerned for my well-being, my parents took me back to the doctor. They were worried the NDE had changed my state-of-mind forever. They wanted the old me, back again. They feared the way I was acting, that I would never be the same again. The doctor told them not to worry.

In the end, they were both right. My consciousness remained changed—and has remained changed ever since. But it is all right. It was meant to be.

My Near-Death-Experience altered the rest of my life. Some have said I was fated to experience it. Regardless, this experience gave me an opportunity to gaze beyond the veil of this physical life into the mysterious realm of the "afterlife" and interact with those who exist there.

I am extremely grateful for the experience. I would call it a "miracle" except my "guides" would disapprove. They know the science of our existence. We do not.

In fact, we have not a clue. That is, I believe, one of the main reasons they want this book written. Our world is out-of-balance. Our gross-materialism; our environmental, political and social chaos is endangering this planet.

My "guides" have determined man's religions are not the answer. Religion has, in fact, held humanity back, as shall be explained shortly.

My past-life dreams began shortly thereafter. Since then, these dreams and visions have both challenged and inspired my life. As you shall discover later in this book, these past-life dreams changed the course of my present life.

CHAPTER 2

The "Christian" Heresy

Life is full of uncertainty. Fate plays its role, but our own free will remains a force—both for better and for worse. But one thing is certain to me. Open your mind to your past-life memories and your present life will never be the same. Reincarnation is a reality.

But most Americans, and Western civilization, have both a mental and an emotional block against Reincarnation. There is a simple reason for this—Christian upbringing (AKA brainwashing). Our reluctance to learn about Reincarnation is the result of almost 1700 years of Christian dogma created by men obsessed with power and greed.

I know. Like millions of others, I was born into a faithful Christian family obediently trying to live according to Christian principles set forth in the Nicene Creed. My parents wanted me to be a pastor and I diligently sought to make myself worthy of such a mission.

There was only one problem. Every time I came close to making a spiritual commitment to pursuing such a pastoral career, I remembered my Near-Death-Experience and the lessons it taught me. Do not believe blindly in unproven fables and myths. Do not follow hypocrisy and lies.

The logical mind cannot accept at face value the validity of myths, legends, folktales, and fables. As much as I tried, even my childhood intellect rejected the notion of an immaculate conception. If Christ is of God, and if God created the universe, then the laws of nature HE created for the entire universe would not allow an immaculate conception for one birth.

It is said that God is truth. Therefore, God cannot be a hypocrite. If God is all about truth, then he cannot violate his own rules. Even though a child, I could figure that out. Plus, whenever I recalled my Near-Death-Encounter, I remembered how obviously unreligious my "guides" were. There overt nonreligiousness resonated with me that there was a higher reality of which I was unaware.

Then, as I grew older, I became increasingly aware of Christianity's incredible hypocrisy, immorality, greed and corruption.

"Beware of false prophets who come to you in sheep's clothing, but inwardly they are ravenous wolves. You will know them by their fruits."

Matthew 7:15-16. NKJB

What are the "fruits" of the "Christian Church?" First, let's separate the "wolves" or Christian clergy from the "sheep" or the faithful masses the "wolves" have led astray.

Was it the faithful who created the Nicene Creed's (as we shall see shortly) restrictions on who went to heaven and who went to hell? Was it the sheep who came up with the clever scheme of selling the faithful "indulgences" so the Church's "wolves" could get rich off the faithful's fear of hell?

Was it the faithful "sheep" who started the Crusades? Was it the faithful "sheep" who started and controlled the Inquisition? Was it the "sheep" that stormed the New World shores of South America with conquistadores proclaiming a cross in one hand while raping, pillaging, and plundering with the other?

Is it the "sheep" who have amassed ungodly hoards of wealth in the name of .....what? What exactly, is the spiritual justification for such wealth? How can a religious organization claim spiritual validity and presume godly sanctity and authority when it has amassed a fortune when billions of people around the world remain poor, undereducated, persecuted by repressive tyrants and diseased?

The Roman Catholic Church is the world's wealthiest religious organization and one of the world's wealthiest institutions. According to the findings of *The National Post*:

"It is impossible to calculate the wealth of the Roman Catholic Church. In truth, the church itself likely could not answer that question, even if it wanted to. Its. investments and spending are kept secret. Its real estate and art have not been properly evaluated, since the church would never sell them. There is no doubt, however, that between the church's priceless art, land, gold, and investments across the globe, it is one of the wealthiest institutions on Earth."

Source: Kristopher Morrison, "Wealth of Roman Catholic Church impossible to calculate," National Post, March, 2013 at nationalpost.com.

The Vatican alone possesses priceless works of art and architecture: including the Apostolic Palace; the Sistine Chapel with its murals by Michelangelo and Botticelli; the painting of the Last Judgment behind the altar; over 716,000 square kilometers of land around the world, including the Vatican monasteries, churches, cathedrals, convents, embassies, and schools; its holdings and investments in banking, insurance, chemicals, steel, construction, and real estate. Its wealth also includes approximately $86 million in member donations per year; one metric ton of gold owned by the Holy See and stored in the Vatican bank; and $308 million in revenue the church received in 2013 alone.

But $31 million was seized by the Italian government during its investigation of Vatican money laundering and an additional $3 billion has been paid out by the Catholic Church in settlements for Catholic priest-sex-abuse-scandal allegations in the United States alone.

Source: Kristopher Morrison, "Wealth of Roman Catholic Church impossible to calculate," National Post, March, 2013 at nationalpost.com. and "Clergy Sexual Abuse Lawsuits," *Miller and Zois, Attorneys at Law* at millerandzois.com. (no date).

"Since 313 A.D., when Catholicism became the official religion of the Roman Empire, its power has been near-constant growth. The church was able to acquire land, most notably the Papal States surround Rome, convert pagan temples, and claim relics for itself. Over 300 years, it became one of Europe's largest landowners.

For the next thousand years, tithes and tributes flowed in from all over Europe. Non-Christians and even fellow Christians were killed and their property confiscated. For example, the Fourth Crusade and the sack of Constantinople in the early 13th Century brought it gold, money, and jewels."

Source: "Wealth of Roman Catholic Church," see above at nationalpost.com.

The price of religious wealth has a negative impact upon family finances. Religious wealth can be tied to institutionalized poverty, according to a study by a Duke University researcher:

"The results demonstrate that religion affects wealth *indirectly* through educational attainment, fertility, and female labor force participation. The results also provide evidence of a direct effect of religion on wealth. Low rates of asset accumulation and unique economic values combine to reduce CP (conservative Protestant) wealth beyond the effects of demographics."

Source: Lisa A. Keister, Duke University, "Conservative Protestants and Wealth: How Religion Perpetuates Asset Poverty," *American Journal of Sociology*, at journals.uchicago.edu, September, 2007.

As would be expected, this ungodly accumulation of wealth has, over the centuries, resulted in a continuing parade of malevolent popes whose warmongering, immorality, murder, torture, and political ruthlessness would make any modern tyrant envious. As much as the Catholic Church would like it, the names of such evil popes as Alexander VI, Stephen VI, John XII, Urban VI, Benedict IX, Leo X, and Boniface VIII will never be forgotten.

Source: Ishaan Tharoor, "7 wicked popes, and the terrible things they did," The Washington Post, September 24, 2015, at washingtonpost.com.

These evildoers are the "trees" from which the "unchristian" Catholic organization offers its foul "fruit" to its pure-of-heart but naïve masses. Protestantism is likewise a corrupted "tree." Do not forget that the Salem witch trials were Protestant inspired.

Please note that Judaism, Islam, the Church of Latter-Day Saints (Mormons), the Church of Scientology and Eastern religions and philosophies do not adhere to the Nicene Creed and do not force their members to accept the conditions set forth in the Nicene Creed.

As desperate as I was to make my parents proud, I could not bring myself to consider teaching and preaching what I believed then—and now—to be a lie anointed in corruption and sanctified by hypocrisy.

Ironically, the only validity of the "Christian Church" is the inherent, spiritual goodness of the people it brainwashed into a blind, obedient following. For this to happen, there must have been some fundamental purity among the faithful which transcended the lies of its doctrine and the corruption, amorality and violence of its history.

Souls are inherently neutral…neither good nor evil. Original sin does not exist. It never has existed except as a tool to rob the people of their wealth and their faith in their own spiritual oneness with the Universal Mind.

It is apparent that the Christian religion plays by two sets of rules. The humble faithful are told they are sinners who must pay the Church in money and obedience to earn the right to go to heaven. Meanwhile, over the centuries, the Church hierarchy has enjoyed living like royalty. Their lust, greed, and corruption hidden from the faithful masses by veils of false-piety, feigned-reverence, and fictitious-authority.

Christian hierarchy have been as bold as they have been deceitful. Christian dogma is to be dogmatically obeyed—even if based upon a series of unprovable myths and undocumented fables. If you want to be a Christian, you cannot believe in Reincarnation. Period. You are forbidden.

Who says so? The Christian Church says so. Both the Catholic and Protestant sects of the Christian Empire (what is left of the Holy Roman Empire of Emperor Constantine) follow the same rules of Christian faith—the Nicene Creed.

Were these "rules of Christian faith set forth by Jesus? No, they were not. There is no evidence of "Jesus" excluding anyone from his father's kingdom.

History records that Roman Emperor Constantine made the Christian faith legal and protected from persecution with his Edict of Milan in 313 A.D. This was a time in which the Roman empire was divided into three sub-empires. By legalizing the Christian faith Constantine consolidated his popularity with a growing segment of the Roman populace, giving him much needed political support during a time when he was engaged in an all-out civil war with the other two Roman Emperors he shared power with.

But Constantine did not legalize Christianity without expecting something in return. The bargain or *quid pro quo* Constantine received in exchange for "freeing" Christians from persecution was obedience. The Christian elders were to submit to his authority and his interpretation of what the newly established Christian organization would become. Emperor Constantine had a plan.

His first step was to create a pope. The pope was intended to be a mere figure-head. Constantine had no intention of allowing the religion he just created to be ruled by anyone other than himself.

The second step was to create the legislative framework by which Christianity would exist and the rules its followers would be required to obey. Constantine meant for the Council of Nicaea to be his "Constitutional Convention." And he chose a Vladimir Putin-type to be his human instrument of authoritarianism—Athanasius of Alexandria.

Athanasius, of Egyptian heritage and birth, was born in 296 A.D. As a young boy, Athanasius was taken by the then Bishop of Alexandria (Bishop Alexander) as a servantboy and became the bishop's personal secretary—even as a young boy. While no proof of an illicit relationship is expressly mentioned, the sad implications are all too apparent, given the tragic history of some senior Catholic clergy.

Source: The Encyclopedia Britannica, Vol. II, (9th ed.), Charles Scribner's Sons, New York, 1878, at ccel.org.

At the time of the Council of Nicaea, he was still a young man, but had already maneuvered himself into the powerful position of Deacon and personal assistant to Alexander, the Bishop of Alexandria.

Athanasius would eventually become Bishop of Alexandria. He would also have to flee for his life from allegations of murder and sexually violating a virgin. He would spend 18 of the next 43 years in hiding from authorities. It has been argued that Athanasius' methods to obtain position and power were questionable at the least and, at worst, was the result of violence.

Sources: "Athanasius of Alexandria," *The New World Encyclopedia*, at worldencyclopedia.org.; Richard Rubenstein, *When Jesus Became God: The Epic Fight over Christ's Divinity in the Last Days of Rome,* New York, Harcourt Brace and Co., 1999 and J. Steven Davis, "The Importance of Athanasius and the Views of His Character," Liberty University, 2017, @digitalcommons.liberty.edu.

Athanasius had two goals when he arrived at the Council of Nicaea. The first was to make sure that only the books he approved of and had edited himself were accepted by the Council as the New Testament. Second, he intended to impose upon the bishops attending the Council an interpretation of Christianity he shared with Emperor Constantine. An interpretation that would become known ever since as the Nicene Creed.

Through force of personality, threats and intimidation, Athanasius and his followers succeeded on both counts. The Council adopted Athanasius' chosen list of twenty-seven books to be included as the sole canon for the New Testament—and excluded everything else. Below is a partial list of the writings Athanasius successfully suppressed from the New Testament:

1. Dialogue of the Savior
2. Gospel of Mary (Magdalene)
3. Gospel of Philip
4. The Sophia of Jesus Christ
5. The Gospel of Thomas

Source: "New Testament Apocrypha," wikipedia.org.

Let's just stay with "The Gospel of Thomas." The original editor of this book is claimed by Syrian Christians to be Judas Thomas, or "Judas the twin," the twin brother of the man we have been taught was named Jesus. This Gospel is very different from the gospels of Matthew, Mark, Luke and John Athanasius made canon.

"Unlike the New Testament gospels…the Gospel of Thomas contains almost no narrative. Jesus in Thomas performs no physical miracles, reveals no fulfillment of prophesy, announces apocalyptic kingdom about to disrupt the world order, dies for no one's sins, and does not rise from the dead-on Easter Sunday. His value, rather, lies in his enigmatic sayings, which are pregnant with possibility and power. 'Whoever discovers what these sayings mean will not taste death.' Jesus promises."

Source: Introduction to "The Gospel of Thomas," *The Gnostic Bible*, page 43, W. Barnstone and M. Meyer, eds., pub. by New Seeds, London, 2003.

Interestingly, whoever Thomas was, he knew Jesus' real name. It is on page 44 of *The Gnostic Bible* (cited above.)

The Hebrew name for Jesus is Yeshua ben-Yosef. If the author of the "Gospel of Thomas" knew "Yeshua" was Jesus' true name, certainly Athanasius and the rest of the bishops attending the Council of Nicaea knew what Jesus' real name was. So why would they intentionally give the man they allegedly worshipped a false name?

The answer is iconoclasm. Iconoclasm is the belief that one's religious or political power and influence can be destroyed by destroying one's image, statue writings or other physical evidence of their existence. In ancient Egypt, thousands of years before Athanasius was born, the ruler Akhenaten sought to destroy the traditional multi-god religion and create a new religion based upon the single god of Aten, the Sun god.

"To reiterate the power of Aten, Akhenaten sent armies all over Egypt to break into tombs and remove Amun-Ra, the original head god under polytheism. As a result, Egypt was "cleansed" of any unwanted images of the ex-sun god strengthening Akhenaten's regime...As a whole, it can be understood that Ancient Egyptian iconoclasm revolved around political agendas, which sometimes bled into the realm of the religious like the case of Akhenaten's image being destroyed

Source: Deogracias, Micaela, "Iconoclasm in Ancient Egypt," Africana Studies Student Research Conference, Bowling Green State University, 2018.

Being Egyptian, Athanasius was obviously aware of the long history of Egyptian iconoclasm. It was part of his culture and that of his own Egyptian people. Creating a fake Bible had the effect of erasing the real Yeshua ben-Yosef from history—along with everything he really taught. Ironical isn't it that the leader of this institutional deletion of Yeshua and what he really believed was, himself, an Egyptian.

If you ask why Americans have a serious problem accepting Reincarnation, the answer is one man—Athanasius of Alexander. Billions of people around the world believe in Reincarnation. They build their lives around it. But Athanasius' Nicene Creed effectively banned Reincarnation from Christian doctrine in 325 A.D.

Ever since, for over 17 centuries, Christians have been conditioned not to believe in Reincarnation. In fact, according to early Christian teachings, if you believed in Reincarnation you were going to hell. If you caught teaching Reincarnation during the Inquisition, you would be burned at the stake.

Why?

Do you go to Christian church? If you do, you must recite the Nicene Creed. Do you believe in the Nicene Creed? Do you know what it says? Do you know what it means? You should, because your priest or pastor has you recite it at the beginning of every service.

Just like the United States Pledge of Allegiance that all American school children must recite every day before class, the Nicene Creed is Christian mind-control in action. This is indoctrination in its most basic form. Let's see what the Nicene Creed says today…

MODERN NICENE CREED

"We believe in one God, the Father, the Almighty, maker of all that is, seen and unseen. We believe in one Lord, Jesus Christ, the only Son of God, eternally begotten of the Father, God from God, Light from Light, true God from true God, begotten, not made, consubstantial of one Being with the Father. Through him all things were made. For us men and for our salvation, he came from heaven, and by the Holy Spirit was incarnate of the Virgin Mary, and became man. For our sake he was crucified under Pontius Pilate; he suffered death and was buried. On the third day he rose again in accordance with the Scriptures; he ascended into heaven and is seated at the right hand of the Father. He will come again in glory to judge the living and the dead. And his kingdom will have no end.

We believe in the Holy Spirit, the Lord, the giver of life, who proceeds from the Father and the Son. With the Father and the Son he is worshipped and glorified. He has spoken through the Prophets.

We believe in holy catholic and apostolic Church. We acknowledge one baptism for the forgiveness of sins. We look for the resurrection of the dead and the life of world to come.

Amen."

Source: "The Nicene Creed," Catholic Online at catholic.org. 2022.

This is the version of the Nicene Creed recited in Catholic churches around the world today. The Protestant version is very similar. On the surface, it seems rather benign. It is not the original Nicene Creed, but one edited over time to make Christian dogma more palatable to modern minds. It is not the same Creed the Christian Church adopted at the Council of Nicaea.

What the Council of Nicaea adopted originally is much more sinister, much more oppressive and much more tyrannical. What you are about to read is what your ancestors were taught as absolute law of the Christian Church from the 4th to the 20th Century…

THE ORIGINAL NICENE CREED (Athanasius) 425 A.D.

1. Whosoever will be saved, before all things it is necessary that he hold the Catholic Faith.
2. Which Faith except everyone do keep whole and undefiled, without doubt will perish everlastingly.
3. And the Catholic Faith is this: That we worship one God in Trinity, and Trinity in Unity;
4. Neither confounding the persons nor dividing the substance.
5. For there is one person of the Father, another of the Son, and another of the Holy Spirit.
6. But the Godhead of the Father, of the Son, and of the Holy Spirit is all one, the glory equal, the majesty coeternal.
7. Such as the Father is, such is the Son, and such is the Holy Spirit.
8. The Father uncreated, the Son uncreated, and the Holy Spirit uncreated.
9. The Father incomprehensible, the Son incomprehensible, the Holy Spirit incomprehensible.
10. The Father eternal, the Son eternal, and the Holy Spirit eternal.
11. And yet they are not three eternals but one eternal.
12. As also there are not three uncreated nor three incomprehensible, but one uncreated and one incomprehensible.
13. So likewise the Father is almighty, the Son almighty, and the Holy Spirit almighty.
14. So the Father is God, the Son is God and the Holy Spirit is God.
15. And yet they are not three Gods but one God.
16. And yet they are not three Gods, but one God.
17. So likewise the Father is Lord, the Son Lord and the Holy Spirit Lord.
18. And yet they are not three Lords but one Lord.
19. For like as we are compelled by the Christian verity to acknowledge every Person by himself to be God and Lord;
20. So are we forbidden by the Catholic Religion to say: There are three Gods or three Lords.
21. The Father is made of none, neither created nor begotten.
22. The Son is of the Father alone; not made nor created but begotten.
23. The Holy Spirit is of the Father and of the Son; neither made, nor created, nor begotten, but proceeding.
24. So there is one Father, not three Fathers; one Son, not three Sons; one Holy Spirit, not three Holy Spirits.
25. And in this Trinity none is before or after another; none is greater or less than another.

26. But the whole three person are coeternal, and coequal.

27. So that in all things, as aforesaid, the Unity in Trinity and the Trinity in Unity is to be worshipped.

28. He therefore that will be saved must thus think of the Trinity.

29. Furthermore, it is necessary to everlasting salvation that he also believe rightly the incarnation of our Lord Jesus Christ.

30. For the right faith is that we believe and confess that Lord Jesus Christ, the Son of God is God and man.

31. God of the substance of the Father, begotten before the worlds; and man of substance of His mother, born in the world.

32. Perfect God and perfect man, of a reasonable soul and human flesh subsisting.

33. Equal to the Father as touching His Godhead, and inferior to the Father as touching his manhood.

34. Who, although He is God and man, yet He is not two, but one Christ.

35. One, not by conversion of the Godhead into flesh, but by taking of that manhood into God.

36. One altogether, not by confusion of substance, but by unity of person.

37. For as the reasonable soul and flesh of one man, so God and man is one Christ.

38. Who suffered for our salvation, descended into hell, rose again the third day from the dead.

39. He ascended into heaven. He sits on the right hand of the Father, God, Almighty.

40. From thence He shall come to judge the quick and the dead.

41. At whose coming all men shall rise again with their bodies.

42. And shall give account of their own works.

43. And they that have done good shall go into life everlasting and they that have done evil into everlasting fire.

44. This is the Catholic Faith, which except a man believe faithfully and firmly, he cannot be saved.

Source: "The Athanasian Creed," The Catholic Encyclopedia, at newadvent.org., 2022.

Though no longer recited in church in its original form, The Athanasian Creed was the genesis for all Christian faith for the last 1700 years and remains the hidden, secret soul of all Christianity. It's format, structure, and content reveal the state of mind of its author, Athanasius. Its structure is mental indoctrination by repetition. It replaces factual evidence with coercion and intimidation and reflects the psychological issues of a sexually-obsessed and psychologically-impaired man. Its meanderings have more to do with psychiatric pathology than theology.

The overbearing use of forced, repetitive phrases are meant to intellectually bully the audience into acceptance despite the fact that Athanasius' writing lacks any semblance of logical foundation or factual support for his ramblings.

This creed of hate, exclusion, spiritual abuse and divisiveness is starkly anti-Christ-like. In fact, no rational person could ever seriously imagine the real Yeshua saying any of the words Athanasius wrote.

Athanasius and his followers fought viciously against the idea that we could be reborn into a new body and live multiple lives. According to Athanasius' creed, we live but one life in the flesh. According to him, when we die, our souls stay with our dead bodies until Jesus (he does not use "Yeshua") returns to Earth.

This "one-life" creed and the threat of eternal damnation to nonbelievers creates both a mental and emotional trauma. When we accept the Christian faith we accept the lie that Reincarnation is wrong. By doing so, we become our own mental jailers—we create our own mental imprisonment that keeps us from exploring Reincarnation objectively.

But this intellectual barrier to Reincarnation is not the only reason we Westerners are reluctant to open our minds up to our past-life memories. Let's pretend that I can make it so you can relive one of your past-life memories right now.

Would you do it?

I think most Americans—and other Westerners, as well—would say, "No thanks." The reason for this response is visceral—fear of reliving a past death.

I don't blame you.

The reason is simple. The transfer of the Soul from one life to the next is a precious moment of life-reaffirmation. But the physical act of dying is not pleasant in the best of circumstances. Quite understandably, we subconsciously resist opening our minds to past-life dreams out of fear of reliving a past death.

I can only speak to my own experience with past-life dreams. Be that as it may, let me share this. I did not begin to have past-life dreams of past-life deaths until I "asked" to have them. By "ask," I mean during meditation to affirm my intention to consciously remember my past-life deaths. As a result, it is my belief that we only recall those past-life events we can cope with emotionally.

Our long-term memories become emotion-encoded by the part of the brain called the limbic system—particularly the amygdala and hippocampus. The more intense the emotional experience, the stronger the limbic system impresses an emotion upon the synapse containing the memory. This is done by varying the amount of biochemically-created electrical charge associated with each synapse. This bio-electric charge is intensified each time we recall the event—making the emotional imprint ever stronger.

Neural recall is conditioned upon emotional cues. Past-life memories are stored deep in long-term memory. Every experience in long-term memory bears an emotional imprint of varying strengths, depending upon the intensity of the emotion you felt at the time the memory was created. Each long-term memory has a corresponding emotional trigger that will open that memory up to you in your consciousness.

The Reincarnation process selects only those memories with the strongest emotional imprint for preservation. This is logical.

Each neuron-synapse contains a memory. Each time you recall that memory the neuron-synapse containing it fires an electrical signal. The more a neuron-synapse fires, the stronger it becomes.

Therefore, the strongest neurons are those containing the memories your consider the most important to your life. These are the memories which have the strongest emotions attached to them. They mean more to you than any other memories. These memories become your "unique individuality"—the memories that distinguish your self-conception as distinguished from everyone else.

These emotion-imprinted memories create your "unique individuality"—your self-concept, your self-identity you carry with you as you exist day-today in your life. This "unique individuality" is what the Reincarnation process preserves when you die by transferring those memories from your dying body into the fetus of one yet to be born.

In essence, your "unique individuality" is your soul. You carry it with you from life to life to life. It is what makes you—you.

Ultimately, what makes our individuality "unique" are those events which shape us the most on an emotional level. No two people have the same emotional experience.

Each of our personalities is unique. We each contain unique information within our memories which differentiate us from everyone else. This is the fundamental causal agent for Reincarnation.

Reincarnation is a function of the limbic system and the central nervous system. The limbic system functions in two ways. When we experience either trauma or joy, we evoke within our brain a biochemical signal which the limbic system receives. The limbic system then imprints an emotionally-laden code upon the synapse storing that memory.

The more intense the emotional experience, the stronger the biochemical signal imprinted within the synapse by the limbic system. The more we remember an event, the more we use the synapse associated with a particular memory, the bigger that synapse becomes.

When you die, the brain searches for the largest and strongest synapses for those memories to transfer to your next body. The stronger and bigger the synapse-stored memory, the more likely that memory will be one the memories chosen to follow you from life to life. As we advance from life to life to life, these surviving memories form a "core of identity" which become you. This is your unique individuality. This is your soul.

This is not a religious miracle. This is not religion at all. This is simply how evolution has adapted your body to meet the natural law demands of the Universal Mind. When we die, the limbic system selects the strongest memories to be transferred to our next body.

In this way, the limbic system serves as a "memory-filter" of out-going memories. When we are born into that new body, those strong emotions are again filtered by the limbic system coming back into the consciousness in our next life as dreams.

As a result, the limbic system blocks any past-life memories that would cause us undue emotional trauma in the next life. We only recall past-life memories that the limbic system has approved for us to remember.

This is also why we never remember our past-life names or where we lived or what our day-to-day routines were in past lives. We never give any of these memories any emotional weight. They are too routine, too common place for the limbic system to preserve them. They have no relevance in your future life.

Look at it this way. Does your unique individuality depend upon what name you have? Of course, it doesn't. Our proper names are just labels, like the street addresses and phone numbers we are given and change every time we move. Labels do not usually evoke a strong emotional response.

Images, however, do.

What we see, we feel. What we feel, we remember. What we remember strongly, we take with us from life to life. The rest is flotsam—mere debris, forgotten in time. Remember that the next time someone tries to dispute Reincarnation claiming if we had past-lives we would all remember our past-life names and where we lived.

The next time someone questions Reincarnation on the grounds you should remember your past life name, ask them to do the same from this life. Have them recite street addresses, zip codes and phone numbers from everywhere they've lived; have them tell you the first and last names of all their old teachers and the street addresses from all their old schools; their first and last names of all their classmates; the street addresses, phone numbers and full names of all their former employers, supervisors and coworkers.

When they admit they cannot recall all of them, explain to them that the reason they cannot is because their limbic system failed to encode any of those memories with an emotional imprint. None of this information was emotionally important to them.

The names, addresses, and other information people do recall are those memories people have an emotional imprint about. We may not recall most of the names of my former teachers and professors, but most of us recall who lives at 1600 Pennsylvania Avenue in Washington, D. C. We may not recall most of our past employers, but those we do remember involved events which created strong emotions. We may not remember all the doctors we visited for health care, but who can't recite the name of their favorite sports teams?

Memories are built upon events which inspire strong emotions. These emotion-loaded memories are like eternal photographs. They exist, deep within our minds and are stored from life, to life, to life, deep within the brain of each body we exist within.

Some memories from our past lives are best left alone. For, with every rebirth comes a new death. Death is not an enjoyable experience. It is not one which anyone is happy to revisit. So, it is not surprising that another reason we have a block against the concept of Reincarnation is the consequential possibility that should we open our minds to Reincarnation, we might just have a vision of a past death.

Why should we seek out our past-life dreams? Of what benefit is Reincarnation to us consciously? In other words, does it really matter that we cannot recall our past-lives if there is no benefit to doing so? How is Reincarnation relevant to our lives in this life?

To answer this question, let me ask you one. Let's assume you have traveled a great distance, and I ask you what you experienced and learned on your journey. To answer me, you must use your memory. Your mind responds by sending you images of what you have done that matter to you emotionally the most. It is these memories that will define your response to my questions. It is these images, and the emotions they bring forth, that will create your answer.

Your answer will create an image of you that I judge and my judgment of you will be reflected toward you as you watch my reaction to your answer. Though we are not consciously aware of it, our past-lives and the emotionally-loaded memories those lives created—made each of us who we are today…

Learn from the joyful souls, for they have found the meaning of life.
Learn from the goal-driven souls, for they have found the purpose of life.
Learn from the inspired souls, for they have found the blessings of life.
Learn from the selfless souls, for they have found the Universal Mind.

Past-life traumas, unresolved, cause anger, hostility, and hate in this life—which only exacerbates the harm caused to both you and those you encounter. The more you understand your past-life traumas, the less you will suffer in this life and your future lives. The more you understand your past-life sufferings, mistakes of judgment and struggles; the less angry, hostile, vindictive and hateful you will be in this life. The happier you will become.

It is not surprising that Westerners would question the relevance of Reincarnation. We and our forebearers have been brainwashed since birth that life is terminal. We only have one Earthy life and then we go to heaven or hell forever.

This is a lie. We have been taught this lie for generations. Life is not terminal. Life is life. Death is merely a transition from one state of matter and energy to another. It is not religion. It is the infinite Universe of which we are a part.

Eastern cultures, especially Buddhist cultures, are taught how to deal with past-life memories as well as past-life death experiences. They are taught to look beyond the suffering of death as an unavoidable but necessary door-way to rebirth—and another life in this world. Their culture also teaches the importance of guided and focused meditation on death and dying to prepare themselves for the inevitable day when they die again.

Source: Sogyal Rinpoche, The Tibetan Book of Living and Dying, Rigpa Fellowship and Harper-Collins Publishers, San Francisco, 2002.

For most Easterners, death is not an ultimate end, but one of many doors they must pass through as they exist from life to life. They have learned not to fear death, but to embrace it as a sort of graduation ceremony as they advance from one level of awareness to the next.

All forms of matter transition from one state of existence to another as the surrounding environment causes changes within it. Water transitions from one state (ice) to another (liquid or gas) as sufficient heat is added.

While consciousness is eternal, matter and physicality are not. Consciousness and the eternal information it possesses is contained in matter. Matter is not eternal. Matter is only temporary and decays over time. Consciousness must compensate for the temporary nature of matter.

Decaying matter dies. Death is but a transition from one state of existence to another. It is a transition from this life to the next. This is not about religious reward or punishment, but as a logical consequence of universal natural laws and evolution.

Death is just an event, a passageway, a physical process of which most of us have very little understanding. Once we realize death does not end who we are but is simply a physiological process of transferring our individuality from this life to the next, death loses its power as a force of fear.

Existence, is therefore, not a closed loop in which everything we learn and experience in this life is lost in death. Nor is it a tortured path of self-doubt and fear of judgment.

No, existence is none of these. The universe and its natural laws make our consciousness and its existence a spiral of ever-increasing awareness and energy into the future—from life to life to life—into infinity.

We are caught in a duality of existential experience between consciousness and the physical universe. Consciousness is quantum-entangled. This makes its existence wholistic—a singular entity. As a result, consciousness is everything, everywhere, all and one, at the same time.

On the other hand, the physical universe itself is linear. At the extreme physical end of the existentialistic spectrum, both time and space impose upon our awareness the false perception that everything is separate, isolated, and distant. Just as conmen and tyrants use lies to accomplish their goals, the physical realm lies to us about the nature of existence. Our physical senses tell us one thing about life but our consciousness presents a different reality.

We are caught in the middle of this conflict. Perhaps one of the most important lessons one must learn about life in order to evolve is how to discern the lie.

To learn how to discern the lies we are confronted with by conmen and tyrants—whether they wear the robes of religion, politics, or marketing, we must allow our consciousness to evolve. To avoid the evil of such "pitch-men," one must first be able to uncover their lies. The answer is education and meditation.

Here are a few lessons.

You cannot attempt to negate a reality within your mind without causing your mind to instinctively rebel. Psychologists call this form of internal psychological rebellion, "cognitive dissonance."

This cognitive dissonance results whenever our life experience conflicts with what our minds tell us the truth really must be. Thus, "cognitive dissonance" is mentally analogous to audio-feedback—painful and unhealthy. It is a warning buoy in the ocean of our consciousness.

Commit a lie and you create a spiritual Karma-wheel. A Karma-wheel is an intentional false thought you have created or allowed to exist within your life. This lie is in motion. It moves with you through time and space like a wheel. This motion is linear and since you cannot forget it exists, it becomes an ever-circling, linear gravitational force within your life. Eventually, this Karma-wheel will come around and fate will force you to face your lie from the other side.

CHAPTER 3

The Scientific Theory

In this life, I have been a lawyer working in Florida since 1983. I have tried well over a hundred jury trials, hundreds of nonjury trials, and evidentiary motions critical to trial advocacy before both criminal and civil courts. I have also won appellate cases before Florida's Second District Court of Appeals.

Both criminal and civil jury trials begin with what we lawyers call an "Opening Statement." An Opening Statement is not argument. Argument is only permitted at the end of the trial when the opposing lawyers present their "Closing Arguments" to the jury. The law limits Opening Statements to only an outline of the evidence the lawyer reasonably expects to present to the jury during the trial and to request the jury render its verdict in the client's favor.

Let us begin.

Ladies and gentlemen of the jury....

Though Reincarnation is, with today's limited technology, not yet directly provable scientifically, there is sufficient circumstantial evidence to make a case for its existence as a causal consequence of fundamental universal laws.

This book is a science-based argument of this theory. We shall take what scientists have already discovered by experimentation, observation and theory. Then we will show how these discoveries demonstrate convincing evidence a Universal Mind exists and that

Reincarnation is a function of natural law. Below is a very brief outline of what we shall show in the chapters that follow:

1. The Big Bang is a recurring system that recycles itself. The collapse of a prior universe becomes the Big Bang of its progeny universe. Universal law makes this process infinite. How many universal rebirths there have been previously is unknown. But the "blueprint" for universal rebirth establishes Reincarnation of intelligent life as a causal consequence of the universal laws that have always existed.

2. The "building blocks" of our present universe were hydrogen, helium, lithium, and massless electromagnetic particles we know as photons (of which visible light is only a small part). For the first 380,000 years, the universe was too hot and too gravitationally bound to itself to allow electrons to bind to atoms and photons to travel. The universe existed as an opaque plasma. Light (photons) could not move normally. It was during this period that all the universal laws and relationships between matter and energy were formed.

3. These laws and relationships existed as data much as electrons form bits that are used to create data we used in computers today. The difference is that during the early years of the universe, this information was absorbed by photons which have properties unique unto themselves. These quantum mechanical properties (including quantum entanglement and nonlocality) bound the laws and relationships controlling the universe onto the photons.

4. After about 380,000 years, the universe cooled enough that electrons could bind to atoms and photons could move freely. Lithium, the lightest metal, was forced above the still very hot mix of hydrogen and helium gases to form a wafer-thin skin of electrically-conductive material surrounding the universe like the skin of an inflating balloon.

5. Photons, now freed, began traveling at 300,000 km/sec and quickly collided with the slower-expanding universe. These photons were absorbed by the lithium-layer surrounding the universe. The lithium-skin performed two functions. As a natural conductor, it created electrical connections between the photons absorbed into it. Second, lithium served as a natural battery, providing electrical power to the now-evolving photons.

6. Laden with data, quantum-entangled with each other and possessing a limitless source of electrical energy; the photons within the lithium shell became self-aware—the Universal Mind was reborn (remember, there have been prior universes).

7. All life within the universe has been repeatedly immersed within these self-aware photons for eons. Photons represent the fundamental building blocks of both life and consciousness. The consciousness we all share has been in existence for billions of years.

8. The natural laws of conservation of energy and information prevent information from being destroyed. Life creates information. Intelligent life creates unique information that universal laws do not allow to be destroyed. Unique information (individuality) must remain unique—otherwise the information is destroyed in violation of universal laws.

9. The Universal Mind created two processes to comply with these universal laws.

The first process, evolution, was necessary to create the physiological adaptations necessary for the human brain to connect to the Universal Mind and to transfer each person's unique information (individuality) from one life-body to the next at death.

The second process is Reincarnation. Reincarnation is simply an evolutionary adaptation of human physiology to conserve our unique information (individuality). The physical process of how human memory is preserved from one life to the next is a natural consequence of human brain evolution. This process creates, stores, and conserves the neurally-encoded electrical signals we call long-term memories by transferring them from a dying body into a fetus as a quantum-photonic wave similar to how the universe itself became conscious after the Big Bang.

10. The initial purpose of the Big Bang was to create a universal consciousness including intelligent life among its component parts—the Universal Mind. Life consciousness is separated enough from the Universal Mind to allow for the illusion of free will in a sea of causality, yet inextricably intertwined within it. Its long-term purpose remains open to much speculation. But this much is clear—the universe is growing in self-awareness—and it is beginning to communicate with us on a conscious level as never before.

That's all there is to it. You have it all now. You have just gotten the "nutshell" version of what this entire book is all about.

Why did I "spill all the goodies" at the beginning? To be honest, I didn't "spill" all of them; just enough to convince you that I mean what I say when I tell you this is something you need to know.

Please consider the ten-point outline my "Opening Statement" to you my jury of one. If you quit reading now, you'll never know why what you've just read is true. You shouldn't just take my word for it.

You deserve to know the full truth. You are the reason the Universal Mind exists. You are its focus. You matter.

Why?

You'll just have to read on to find out. Believe me, once you open your mind to the Universal Consciousness, your life will never be the same again.

Why has it taken this long for us to realize how we are connected to the Universal Mind and Reincarnation? For decades now, scientists around the world likely already have concluded this theory must be true. It is the only logical consequence of the circumstantial evidence you are about to read in this book. For educated people with critical thinking, the vast amount of circumstantial evidence leads to no other conclusion.

However, just as military and commercial airline pilots refused to publicly acknowledge their personal encounters with UFOs out of fear they would be grounded or fired; scientists have refused to publicly admit or discuss Reincarnation even among themselves.

The reasons for this unwillingness of scientists to discuss Reincarnation as a science-based natural consequence of our photon-based existence are two-fold:

First, quantum physicists—the academics most familiar with photonic mechanics—only began appreciating the incredible implications of what they are learning about these massless wave-particles called photons within the last twenty years.

Second, like everyone else, American scientists have been raised in a Christian-dominated culture. They have been indoctrinated since childhood in religious teachings which view science as heresy. Even today, American scientists and teachers must walk a legal-tightrope between educational truth and politicalized-religious-dogma whenever they enter a classroom. The medieval Inquisition is always lurking close at hand.

If you doubt university faculty in America are frightened by just how far these neofascist attacks on academic freedom may go, consider the following quote:

“We don’t often post editorials or opinion pieces that aren’t directly about higher education in Florida. But these are not ordinary times, especially in Florida. The attacks on tenure and academic freedom at Florida public universities are part of much larger attacks on the teaching of science, history, race, and slavery, among other things, in Florida schools…Republicans have made considerable political hay by denouncing the teaching of critical race theory; this strategy has succeeded even though most voters have no idea what that theory is and its isn’t actually being taught in public schools. But the facts don’t matter because denunciations of C.R.T. are basically a cover for a much bigger agenda: an attempt to stop schools from teaching anything that makes right-wingers uncomfortable.”

Source: Editorial Opinion of United Faculty of Florida (UFF-FAU), “Attack of the Right-Wing Thought Police,” at uff-fau.org., Jan. 27, 2022

The scientists who would understand the concepts behind the theory of Reincarnation as science: physicists, cosmologists, astronomers, etc., do not want anything to do with politics or ideological warfare. They just want to teach, do research, and publish. Academics are not lawyers. They do want to teach, not litigate. Academics are not soldiers. They want to teach, not fight.

American neofascists are copying the Nazi “playbook” of how to destroy a democratic society. Academic freedom and the freedom of ideas is the greatest threat to a dictator and the first human right tyrants seek to destroy.

Today, neofascists intimidate professors, attack academic freedoms, and chill free speech. Tomorrow, it may very well be you who is attacked. Consider what happened to Albert Einstein in Nazi Germany…

“Inevitably, Einstein’s fame and the great success of his theories created a backlash. The rising Nazi movement found a convenient target in relativity, branding it “Jewish physics” and sponsoring conferences and book burnings to denounce Einstein and his theories. The Nazis enlisted other physicists, including Nobel laureates Philipp Lenard and Johannes Stark, to denounce Einstein. *One Hundred Authors Against Einstein* was published in 1931.

In December 1932, Einstein decided to leave Germany forever (he would never go back). It became obvious to Einstein that his life was in danger. A Nazi organization published a magazine with Einstein's picture and the caption, 'Not Yet Hanged' on the cover."

Source: "Albert Einstein," Encyclopedia Britannica, at britannica.com. (No date).

If the greatest scientist of all time can be targeted by a power-mad maniac like Hitler and his henchmen, is it impossible for it to happen here, in America? Five years ago, would you have imagined a horde of militant neofascists would assault the National Capitol?

There is but one reason "conservative Christians" (code words for "anti-science") fear academic freedom. They fear certain truths. Truth about America's historically institutionalized racism and discrimination makes them uncomfortable. They fear the destruction of institutionalized racism will be the end of their privileged life-styles.

White-privileged lives, built upon the shifting sands of subjective emotions and falsehoods, conflict with objective reality. Their response is not to adapt to the reality but to pass laws making it illegal to acknowledge the reality's existence.

The radical-Right (neofascism and white supremacy) are much like ISIS. They fear change and its militants are willing to show off their love of firearms and violence to threaten and intimidate anyone who dissents. We see how the radical-Right fears Critical Race Theory and Gay-equality.

Ideas threaten the radical-Right. I recall one day giving a high-school friend of mine a copy of Kahlil Gibran's book, *The Prophet*. Published by Alfred A. Knopf, NY,NY, 1923.

In case you have never heard of this amazing work, it is a collection of poems which tell the story of a traveling prophet discussing life. First published in 1923, it has never been out of print. Translated into over 100 languages, it has become one of the world's greatest works of literature of all time.

A few days later, I happened to see my friend in the school hallway. As soon as she saw me, she started crying. She explained that not long after she had gotten home, her father had discovered the book I had given to her.

She told me her dad angrily took the book from her, took it to his Knights of Columbus meeting and made her go with him. She said she was forced to watch as the members of the Knights of Columbus lodge ceremoniously burned the book.

The experience had obviously traumatized her. I have never forgotten her grief that my gift to her met such a fate, nor her shame that her father would succumb to such infantile, irrational fears.

Of course, I did not witness this book-burning ritual. Her testimony is only hearsay. But, if true, it may be the first time an alleged group of "knights" ran in fear from a book. And they called themselves "knights?"

The depths of ignorance breeds uncontrolled mob panic and hysteria. It is this paranoia which caused the Inquisition and the Salem witch trials. It is this nuttier-than-a-fruit-cake mentality which led to Joan of Arc's being burned at the stake.

Imagine how they will react when Reincarnation threatens to overturn Christian myths. Reincarnation as a scientific theory is growing in popularity across America across all age groups. Many are curious but until now, no one has attempted to fully explain how the Universal Mind and consciousness are inextricably linked to Reincarnation—and in fact make Reincarnation a causal consequence of natural law.

Yes, Reincarnation threatens traditional myths upon which Christianity is fabricated. This is because Christianity, as we know of it, was created to destroy what Yeshua taught.

But Reincarnation is not a threat to spirituality. It never will be. Spirituality is the relationship between us and the Universal Mind. That relationship will never be at risk because that relationship is intrinsic to our existence.

If scientists cannot explain to the average person the methods and techniques of their specialty, then expecting the public to understand the relevance of science to their daily lives becomes increasingly difficult. We are already suffering from the effects of this "education-gap" in the political resistance the fascist, radical-Conservatives have been able to mobilize against efforts to reduce the impact of climate-change and to mandate anti-Covid vaccinations.

People have a natural distrust for anything they do not understand. The methods and techniques used by modern scientists to study and document climate-change have become too sophisticated for the average person to understand. This is a significant problem, and it is getting worse. Even today, 61 per-cent of the American population still lack a college degree.

Source: Melanie Hanson, Education Attainment Statistics, educationdata.org, 2021.

College educations are necessary for an individual to develop critical thinking and analysis capabilities. College is also helpful in keeping up mentally with the ever-expanding scientific discoveries and technological advances that our society is based upon.

Please do not confuse "career-schools" and post-high-school "technical schools" with traditional liberal arts colleges. Many students today fall prey to the ploy used by "career-schools" that there is no need to take up four years of their lives when all they need is a few months learning how to do "x-y-z" as a truck driver, medical assistant, legal assistant, law enforcement officer or guard, etc.

Those months spent learning a trade do not teach who you are and what choices you will make and how you will respond should you face a critical moment that will forever define your life and those who were lucky or unlucky enough to encounter you in that critical moment.

For example, what would you do if someone had asked you to join the January 6th, 2021 alleged riot-insurrection at the National Capitol building?

Remember, what happened at our nation's Capitol building January 6, 2021 was very serious. In addition to attacking our democratic institutions, this alleged riot-insurrection resulted in injuries and the loss of human life. Under Florida law the loss of life could have resulted in criminal charges for felony murder, culpable negligence homicide and/or manslaughter.

Would you join this alleged riot-insurrection or would you call the FBI to report that alleged militants were allegedly planning to attack the Capitol building? Your answer to that question reveals a great deal about your common sense, your respect for our national democratic institutions and the rights of others. In essence, your answers reveal to what extent you are capable of living in a civilized society of laws we all must obey.

College is not just about getting a degree and a career-path. College is meant to teach how to think independently and objectively to make sound judgments in life.

So, before you spend your hard-earned money on a "quicky" career-centered school, ask yourself the following:

1. Is this school going to teach me how to think for myself so I don't accept as truth any lies my future employers, co-workers or customers may tell me?

2. Is this school going to help me tell the difference between truth and fake-news?

3. Is this school going to help me make the decisions I need to make to keep from making a wrong choice that might ruin my hard-sought career?

If your college does not help you accomplish these three tasks, why are you wasting your time, money, and your own mind on such garbage? A liberal arts education does more than give you a piece of paper you can use to get a job and, hopefully, the birth of a fulfilling and enriching career. A liberal arts college teaches you the value of critical and analytical thinking to enrich your life.

If you want to be your own boss in life, you first must be the "boss" of your own mind. If you can't think for yourself; if you must rely upon others for the choices you will be held accountable for; you're just a zombie being led by the nose to self-destruction.

There is a price we all pay for the increasing numbers of people in American who lack the ability to think for themselves. As science and technology continues to rush forward at light speed, too many are left behind due to their inability to comprehend the advances being made and how those advances are relevant to their routine, increasingly irrelevant day-to-day lives.

This "communication-gap" makes these uneducated "Jane and John Does" easy prey for right-wing, radical-conservative fascist demagogues who seek to manipulate the uneducated into a mindless mob willing to pursue whatever their fascist leaders want done.

The only cure is education. This book is an effort to do just that. I realize that this book is not the easiest to read. Some parts of this book are difficult because of the science. Believe me, if there was some easier way to explain the science to you, I would.

But unless I use the science upon which the theory is based, you would be left with only my word to go on whether to believe the theory. I cannot have you rely upon only my word. You must understand this for yourself. After all, it is your soul we are trying to help here.

What must be done is try to explain the unexplainable in such a way that the average person, uneducated in quantum mechanics, can understand. The assumption I have made is that the average person is not stupid, just uneducated in the complex mathematical world of quantum mechanics, elementary particle physics and relativistic phenomena that make up our photon-dependent lives. We will try to explain without all the math and hope that is good enough. That is the goal.

Truth matters. The scientific evidence of Reincarnation presented here threatens the very foundation of Christianity, which continues to stubbornly deny its existence.

Salvation? Damnation?

Look at those two words, for a moment. Don't they sound intimidating? Imagine the force these two words take on when evoked from the mouth of a bombastic alpha-male, obsessed with his own false sense of superiority, wearing the robes of artificial authority, standing on top of an elevated altar, high above the fearful congregation.

With Reincarnation, neither of these fear-evoking and obedience compelling words have any meaning. That is why Christians are taught from birth to reject Reincarnation and fear death—that is the only reason for anyone to pay homage to the false Christian and Islamic god of death and judgment.

Salvation? Damnation?

Two mental states created by religion to indoctrinate the masses in fictional narratives which could fool our simple agrarian ancestors but should no longer fool us. All it takes to be freed of their lies is to be able to think critically for ourselves.

The truth of Reincarnation allows everyone to live their lives in peace, freed from religious doctrines, to finally know they evolve from life to life without having to bend their knee to false authority.

Just keep an open mind—and know that you are a freed soul; no longer beholden to religion, its fictions, or antiquated and false "mysteries" Christianity uses to claim its faith is the only valid faith worthy of heaven.

The purpose is to explain to you the scientific principles which make Reincarnation a fundamental part of our lives, so you can understand how it works and why it is a fundamental, causal consequence of the universe. I want you to meet my Maker.

But the truth is that my maker is also yours as well. This book is for you, the average person, who lacks a scientific background. I want you to understand, as much as possible, the scientific bases for the incredible theory you are about to be told in the chapters ahead.

If you are open-minded enough and willing to follow with me as I take you by the hand through this book; it will be more than worth your while. Once you open your mind to your own past-life memories, your present—and future—lives will never be the same.

CHAPTER 4

Reincarnation: It Takes a Universe

Scientists have discovered the basic physics establishing not only that Reincarnation exists but that it is a fundamental consequence of the Big Bang. No, you cannot "prove" or "disprove" Reincarnation any more than you can "prove" or "disprove" quantum mechanics. You won't see academic papers celebrating the discovery in such scholarly publications as the *Physical Review Letters* or *Nature*, or *Nature Physics* or the *Journal of High Energy Physics* or the *Monthly Notices of the Royal Astronomical Society* or the *International Journal of Theoretical Physics*.

Reincarnation is too controversial for such publications to accept papers on such a radical theory. Imagine if Albert Einstein or Stephen Hawking had proclaimed Reincarnation existed. Such an announcement would send stock markets into panic-selling frenzies. Pulpits in churches around the nation would melt under the heat of ministerial vitriol. Would religious persecution of American academics not be far behind?

Doubt me? Think you're capable of dealing with a theory and the scientific knowledge in support of it that will change your life?

As we will discuss in detail in following chapters, the universe began with a Big Bang in which hydrogen, helium and lithium atoms were released along with electromagnetic energy—what we call photons.

Two series of photon-waves were emitted at the time of the Big Bang. The first wave was released during the Big Bang itself; the second wave approximately 380,000 years later. Due to the enormous gravitational and chemical instability of the early universe, the first photon emission became trapped within an opaque plasma-field.

Even now, our instruments cannot pierce this opaque plasma-field to view the Big Bang itself. There is a reason we cannot "see" anything closer to the Big Bang than 380,000 years post-Big Bang. For these first 380,000 years, or so, the temperatures and gravitational pressures present within the "plasma-field" encapsulating the new universe were so great that electrons could not bind to atoms and photons were held in place.

Since the photons could not move, no light could emerge. This formed the opaque-looking image we now observe on our telescopes. Because electrons could not bind to atoms, this period of the universe's development is called the "decoupling era" because the electrons could not "couple" or be bound to atoms.

This photon "cloud" encapsulated the entire universe as this plasma-field lasted for approximately 380,000 years. But do not think the universe was not evolving during this time.

The universe was very busy. During this time, all the natural laws and relationships between matter and energy were determined. In this way, the photon-cloud was the broth in which all the universal laws were forged and stirred, like a soup. This "soup" was mixed for approximately 380,000 years.

After approximately 380,000 years, universal expansion caused the temperature and gravitational forces holding light within the plasma-field to lessen significantly; enough that light was able to travel at its normal light-speed. This event released a second massive photon-wave which has since spread throughout the universe like an infinite ocean. It is this second "photon-wave" that we detect today throughout the cosmos.

This photon-cloud that exists throughout the universe is quantum (universal) consciousness. This cloud of photons, with ionized-particle properties has saturated the entire universe since the Big Bang (like the Internet "Cloud"). This cloud, invisible to the human eye, possesses the capacity to share information with that which it comes into contact.

This sharing of knowledge creates "awareness" which evolves with physiological evolution until there is ever-growing sharing between the Universal Mind and the individual. This sharing becomes a form of "quantum entanglement" which, we will learn shortly, allows each individual photon to instantly share its information with every other photon it ever becomes entangled with.

This instantaneous data transfer would appear to violate the photon's speed limit of 300,000 km/sec but quantum mechanics allows for it. Quantum entanglement also explains how the universe became self-aware and how our consciousness, which consists of photons, is connected into the Universal Mind.

Photons are subject to the Conservation Laws of both Information and Energy. These laws were first discovered in the early years of the industrial revolution when steam was the main source of energy for industrial nations—hence the name "thermodynamic."

However, science has evolved significantly since then—and with it our perspective of the Conservation Laws of Thermodynamics. It is now understood that the term "thermodynamic" fails to properly identify all the physical consequences of these laws.

> "The laws of thermodynamics are special laws that sit above the ordinary laws of nature as laws about laws or laws upon which the others depend. It can be successfully shown that without the first and second laws of thermodynamics which express symmetric properties of the world (or universe), there could be no other laws at all."

Source: "All about entropy, the laws of thermodynamics and order from disorder," Archives of Science, 2001, reprinted from entropylaw.com.

While these laws originally were meant to apply to steam engines, the principles underlying these laws apply to any closed system that consists of the following variables: a gas or gases, a vacuum (as a relative change in gas density), heat, pressure (as gravity) and time.

If this reminds you of something other than a steam engine, you are correct. These variables perfectly describe the contents of our universe. Therefore, any consequences arising out of the thermodynamic laws must apply to the universe and any particles that exist within it—in other words—us. The following is a simple description of these laws as they were originally applied:

CONSERVATION LAWS

1. First Law of Thermodynamics (old)

Any thermodynamic system in an equilibrium state possesses a state variable called the internal energy (E). Between any two equilibrium states, the change in internal energy is equal to the difference of the heat transfer into the system and work done by the system.

2. Second Law of Thermodynamics (old)

A natural process that starts in one equilibrium state and ends in another will go in the direction causing the entropy (randomness or unpredictability) of the system plus the environment to increase for any closed (irreversible) system and to remain constant in any open (reversible) system.

Source: "First and Second Laws of Thermodynamics," Glenn Research Center, grc.nasa.gov. 2022.

If that sounds like a bunch of geek-gobbledy-gook to you, do not be concerned. We do have the English translation for each law:

1. First Law of Thermodynamics (new)

"All real-world processes involve transformations of energy, and the total amount of energy is always conserved (is neither created nor destroyed). The universe is a continuum of energy which remains the same regardless of transformation. As to energy, time is symmetrical. The same energy exists for all time, simply changing states over time."

2. Second Law of Thermodynamics (new)

"In all natural processes, the total entropy of a system plus its environment cannot decrease. It can remain constant for a reversible process but must always increase for an irreversible one. The universe acts spontaneously to minimize potential outcomes of any process. Therefore, the universe always seeks to maximize entropy (randomness). As to entropy, time is asymmetrical. Therefore, the Second Law of Thermodynamics creates a measure of the transformation of energy over time."

Source: "All about entropy, the laws of thermodynamics and order from disorder," Archives of Science, 2001, reprinted from entropylaw.com.

Perhaps, we can make these concepts even simpler to wrap our brains around. Let's begin with "entropy." This is very simple. "Entropy" is just a logical consequence of how any system changes over time.

Before a soccer game begins, the state of the game as a system is very simple. The score is zero-zero. The "system" is the game and its rules. The "particles" whose movement will determine the change of state of the system are the players.

We what each player must do before the game starts—the players must take the field. This is easy to predict. If the players do not take the field, there IS no game. But once the teams take the field, predicting where each player will go after that becomes increasingly more difficult as the game progresses.

With each kick, header, throw-in, substitution and goal, the "state of the system" of the game (how each player will react over time within the game) changes dramatically until it is very difficult or impossible to predict what any player will do, due to all the intangible possibilities that arose from all the changes that occurred previously.

This increasing or evolving unpredictability is "entropy." Do not confuse entropy with chaos, disorder or other synonym which implies possible instability within a system. Entropy is different from instability. The closest synonyms to entropy may be "randomness" or "uncertainty."

Entropy has to do with how predicting outcomes of a system becomes increasingly more difficult as time passes.

Why?

Because as time passes, systems change. Systemic change results in more information about that system which must be accounted for to predict future outcomes of the system. Like I said, once you get away from all the gobbled-gook, the concept of entropy or increasing uncertainty is very simple.

Would you believe that our universe is not any different from an ordinary tea kettle? Being a closed system, our universe is subject to the laws of thermodynamics like any other closed system. The universe possesses gases, vacuum (or relative change in gas density), particles, heat, pressure (as gravity) which change over time just like any other closed system.

At the time of the Big Bang, entropy was relatively low. Over time, entropy (uncertainty) increased. This change in entropy, the increasing uncertainty, can be measured. This continuingly changing measurement of uncertainty is information. Over time, measurements of change in the system being observed creates information unique to that system.

Information, change, and time are fundamental components of all systems—including our universe and us. We humans are systems, too. We humans exist as complex biological systems containing subsystems (central nervous system, circulatory system, digestive system, endocrine system, reproductive system, etc.).

This fundamental principle leads us to the logically inevitable conclusion that all knowledge is no more—or less—than measurements we make upon the change in our environment as time passes. Information is the inevitable result of the universe's ever-changing balance between matter and energy over time.

Does this make us no more than tea kettles with a brain, under the control of universal forces we do not now, nor may ever, understand?

Since the Second Law of Thermodynamics provides a measurement of how the universe changes over time, scientists now conclude that this Second Law governs the conservation of information as well as entropy. In fact, it is now scientifically accepted that information can no more be created or destroyed than either matter or energy can be.

Look at the conservation of information like this:

Ever since you've been born, you have witnessed existence—you own and that of all you have observed. This information is saved in memory in neurons and synapses of your brain. Delete those memories, delete that lifetime of information—or any part of it—and you no longer exist. Your unique individuality (all the information you acquired during your life) would be lost—forever.

As stated in the thermodynamic laws, time is asymmetric—irreversible. Just as we cannot ever go back in time, we cannot ever lose the information we gained over that time. So, it is possible to conclude that information (the continued change of states of matter and energy) equals existence.

"Information is physical, it cannot simply disappear in any physical process. This basic principle of information constitutes one of the most important elements for the very formulation of our daily life and our understanding of the universe."

Source: B. Zang and Q. Cai, et al, "Information Conservation is fundamental: recovering lost information in Hawking radiation," arVix, 2013.

The early stages of the universe involved primal forces on an infinite scale. Nevertheless, the forces at work during the early stages of the universe had to ultimately comply with the laws of thermodynamics just as they must do so now. Hydrogen-helium gases and lithium metal reacted under intense gravitational pressures and heat as photons and electrons began to move freely. The universal "tea kettle" may have been of infinite size, but it was still basically a tea kettle.

How did the photons react within that dynamically evolving system?

Photons absorb, carry and release information very easily. The photons at the beginning of the universe were physically part of the process of universal law and relationship formation. Photons absorbed this information and carried it with them until they were absorbed into the lithium-ion skin covering the universe.

Photons are subject to the Second Law of Thermodynamics. Therefore, the information the photons absorbed and carried as the natural laws were being created during the early universe cannot be destroyed. Information cannot be destroyed.

Consciousness is simply awareness of one's own existence. Consciousness is inextricably intertwined with information. Our existence—its objective and subjective realities—is information. Delete this information and you destroy the consciousness that held it in place.

Do you want proof you live forever? There is a very simple logical argument called a syllogism that no religion can refute. No one can dispute a logical premise that is consistent with its conclusion. Consider this:

Major Premise: Information cannot be destroyed.

Minor Premise: Consciousness is information.

Conclusion: Consciousness cannot be destroyed.

No religion will teach you that. But think about it for just a moment. If this conclusion is valid, then any religion promising a "reward" of an eternity in heaven if only you submit to its beliefs and rules is nothing but a fraud. If consciousness is indestructible, we are already eternal, have been, and forever shall be. Human religion has just become irrelevant.

Photons interact with the electromagnetic field of the human brain all the time. This interaction bathes every living organism with photonic information which, at the quantum-level, connects the living organism to the Universal Mind. This intermixing has been going on for millions, if not billions, of years. It likely resulted in constructing the parameters within which evolution determined how the human brain evolved.

The evolutionary parameters were set by the same natural laws absorbed by the photon-plasma after the Big Bang. These parameters set in motion the evolutionary development in human brain and physiology necessary for Reincarnation. Lower life-forms do not have the brain development to perceive the existence of the Universal Mind or be self-aware.

However, through evolution, the human brain eventually became self-aware. This "self-awareness" allows consciousness to be shared between the Universal Mind and the individual human. This sharing increasing as the individual human becomes increasingly self-aware.

Meditation and individual growth create changes to the individual's consciousness-wave-form which deepens the individual's conscious-oneness with the Universal Mind. This growth is information and can be measured as a change of state of the individual's consciousness. Therefore, this increase in awareness cannot be destroyed at death, but is preserved from one life to the next.

Reincarnation is NOT a mystery. Neither is it mystical or even strange. All Reincarnation does is what natural law mandates—conserve information and consciousness. Reincarnation is a logical, necessary consequence of the Big Bang and the natural laws it set in motion thirteen-plus billion years ago.

If you are interested in the specific human brain and physiological development used in Reincarnation, I suggest you read: ***REINCARNATION: The Scientific Theory***, available in ebook and paperback, at most retailers.

This theory of Reincarnation as a natural consequence of the Big Bang is expressed by the formula: $R = uc(ci + ce)c^2$ where "R" stands for Reincarnation; uc stands for universal consciousness; ci for Conservation Law of Information; ce for Conservation Law of Energy; and c^2 for the speed of light, squared.

We are familiar with the Conservation Laws of Information and Energy. They are natural laws determined by the processes involved in the formation of the early universe shortly after the Big Bang. But what is this Universal Mind and where did it come from? The answer begins with the Big Bang, itself.

CHAPTER 5

"The Big Bang"

We are about to take a big step and discuss why photons are so important to the creation of the Universal Mind and our own consciousness. So, to prepare for this, let's review a minute before we go forward.

Approximately 13.7 billion years ago, the universe was created by an unknown force resulting in what we now describe as the "Big Bang." This is what the world's leading cosmologists and physicists believe happened.

The Big Bang released an enormous number of high-energy photons. It is this initial flash of photons that, over time, created consciousness.

Following this initial, enormous flash of photons, the universe existed as a white-hot, primordial fireball of photonic and ionized-particle-plasma. The Big Bang produced only three elements. All matter released directly from the Big Bang consisted of 95 percent hydrogen, 5 percent helium and a trace amount of lithium.

Due to the intense heat and gravitational forces existing within this fireball, electrons were unable to attach (be bound) to atoms and photons (light) were unable to move. For the next 380,000 years, or so, the universe remained in this embryonic plasma-state as all matter and energy were bound together—and all the fundamental laws and relationships governing these two primal ingredients—came into existence.

Modern telescopes are sophisticated enough to look back in time almost to the Big Bang itself. Using infrared, microwave, and ultraviolet recording instruments, these telescopes can detect and record all the way back in time to approximately 380,000 light years following the Big Bang. But that is as far as our modern technology can take us.

Researchers at Harvard University and the Smithsonian Institution explain why we can see no further:

"No stable atoms existed in the early universe, because collisions between particles of light called photons and matter kept knocking electrons away. Those collisions also meant the cosmos was opaque because photons couldn't travel far.

But about 380,000 years after the Big Bang, the expansion of the universe allowed the first stable atoms to form (the universe had cooled enough to allow electrons to stick to molecules). This event is called "recombination" or "decoupling", and it made the universe transparent. The joining of electrons to nuclei produced a lot of light, which could now travel through the less soupy universe and even reach us, 13.8 billion years later."

Source: "Cosmic Microwave Background," Center for Astrophysics, Harvard University and the Smithsonian Institution, cfa.harvard.edu, no author, no date.

NASA scientists have also described this phase of the universe:

"According to theoretical models, shortly after the Big Bang, the universe was filled with a glowing plasma or ionized gas. As the universe cooled and expanded, electrons (negatively charged particles) and protons (positively charged particles) began to bind together to form neutral (uncharged) hydrogen atoms. As the light of the light from the Big Bang spread out from the once-dense mass of the universe—now about 378,000 years old—the universe would have been a dark place, with no new sources of light to illuminate its fog of cooling, neutral hydrogen gas."

"What would the Young Universe Look Like," Green, Thomas, et al, NASA's James Webb Space Telescope Institute, webbtelescope.org, May 31, 2018.

This is the present theory of the Big Bang and its consequential expansion held by scientists around the world: a massive release of photons, followed by consolidation of matter and energy for millions of years until new light emerged from the formation of the first stars.

Merging the Harvard-Smithsonian description of the early universe with that described by NASA scientists, we have the following vision of the early universe:

1. The Big Bang was precipitated by an immense release of photonic radiation, which may have included more than one wave, like a cosmic tsunami. The term "immense release" probably does not come close to the actual amount of light released; and there may be no human terms that adequately approximate the event.

2. The incredible temperatures within the early universe created a plasma-field in which all the laws and relationships affecting matter and energy were determined. Scientists theorize that these super-hot temperatures prevented electrons from binding to atoms, thus holding the initial light within the plasma-field.

It is also suspected that this electron-nonbinding effect would not have been sufficient to hold the first wave of photons within the plasma-field. Therefore, it is theorized that the electron-nonbinding effect was accompanied by an intense gravitational field holding the photons within the plasma-field, much like a black hole.

3. Two processes were at work within the plasma-field. First, this plasma-field was a cosmic-soup where high-energy particles and photons were in proximity with one-another, under high heat and gravitational pressures.

Second, it was during this plasma-stage of the universe's development that all the laws and relationships affecting matter and energy were determined. As we shall see shortly, light is an excellent vehicle for absorbing, carrying, and emitting information.

Due to the immense gravitational force within the plasma-universe, the initial wave (or waves) of photons became bound ("coupled") to the early matter and the information it contained regarding the formation of the natural laws affecting life.

It is theorized that what scientists call "dark matter" and "dark energy" are the remnants of this distant consolidation of matter and the initial photon-waves from the Big Bang. It is believed that "dark matter" cannot interact with our physical reality because "dark matter" lacks electrons. This is consistent with the early universe being so hot that electrons were unable to bind to matter or photons, causing a sort of primordial plasma-soup.

The only energy existing in the early plasma-soup epoch of the universe was photonic radiation—high energy particles heated and squeezed by intense gravitational forces to become bound to matter. Some of this energy was released as part of the great "uncoupling" that occurred after 380,000 years.

But the rest of the photons may have become too bound to the primordial matter to be released during the "decoupling." It is possible these photons became forever trapped as "dark energy." As a result, both the "dark matter" and the "dark energy" bound up within it would be unable to interact with our reality because photons are massless particles with not weight of their own. If they were bound to primordial matter, neither the photons nor the particles they were bound up with would be able to interact with normal matter and energy. This would explain both "dark matter" and "dark energy."

4. After about 380,000 years following the Big Bang, the early universe cooled to the point where electrons were finally able to bind with atoms and photons could "decouple" or be unbound to the matter within the plasma-field.

This "decoupling" created a second sudden, immense flash of photonic radiation that sped away from the plasma field at light speed. It is this second flash of photonic radiation that has swept through the universe, filling it with an ocean of light and information released from the universe's process of natural law formation.

It is this second, "post-decoupling" photon release that carries with it the treasure the universe has given us—a second wave of light that we detect today through what is called Cosmic Microwave Background (CMB) radiation.

Observations of the CMB include NASA's Cosmic Background Explorer (COBE) satellite probe launched in 1989; NASA's Wilkinson Microwave Anisotropy Probe (WMAP) in 2001 and the Planck Space Observatory, launched by the European Space Agency (ESA) in 2009.

Each probe resulted in a more detailed and accurate radio-telescopic image of the universe shortly after the "decoupling" occurred, with the Planck images being the most recent. The Planck images are far more revealing than the images received from COBE or WMAP probes. Further, neither COBE nor WMAP instruments could examine multiple channels of data in the microwave region of the electromagnetic spectrum, as the Planck instruments are able to do, as seen below.

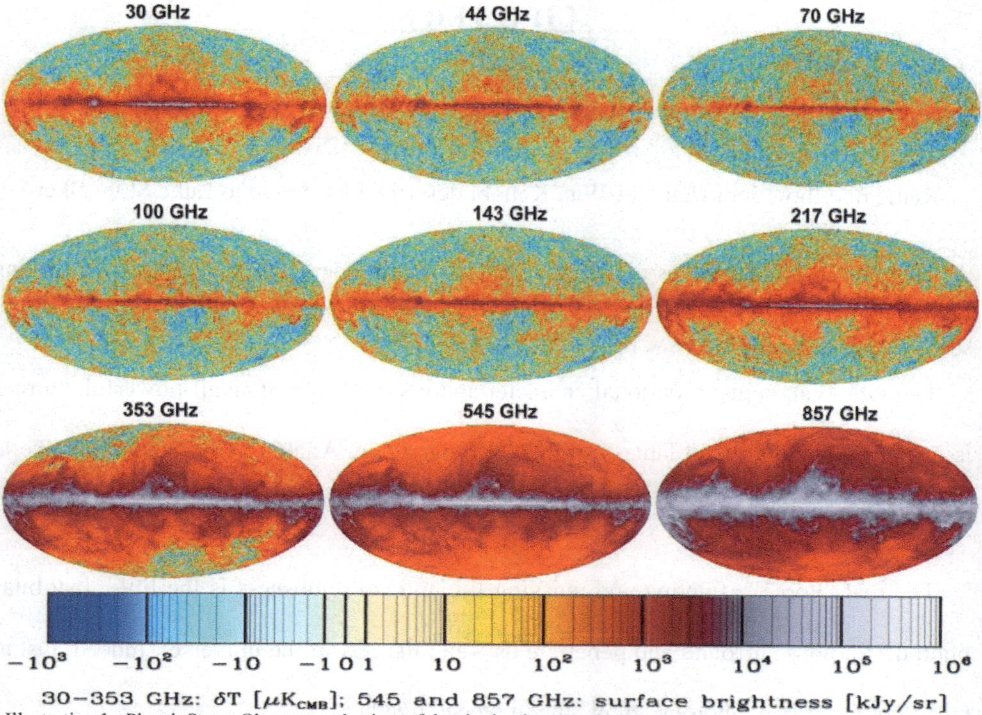

Illustration 1: Planck Space Observatory's view of the sky in nine wavelengths.

Source: ESA and the Planck Collaboration copyright 2022.

"Besides mapping the temperature of the cosmic microwave background across the sky with unprecedented accuracy, Planck (Space Observatory) also measured its polarization, which indicates if light is vibrating in a preferred direction. The polarization of the cosmic microwave background carries an imprint of the last interaction between the radiation (the photonic wave) and matter particles in the early Universe, and as such contains additional, all-important information about the history of the cosmos. But it could also contain information about the very first instants of our Universe and give us clues to understand its birth."

Source: "From An Almost Perfect Universe to the Best of Both Worlds," ESA, @ sci.es.int, July, 2018, no author.

The microwave images above are of the universe, as it existed just after the great "uncoupling" of photons and matter. Now, let's focus on the "decoupling" flash of photons. Imagine a wave of photonic energy flooding through the infant universe at the speed of light.

CHAPTER 6

The Mighty Little Photon

Remember how Jedi Ben Obi-Wan Kenobi described the Force to Luke Skywalker?

"The Force is what gives a Jedi his power. It's an energy field created by all living things. It surrounds us and penetrates us. It binds the universe together."

Source: "Star Wars: Episode IV, A New Hope" 1977.

Jedi Obi-Wan is just a fictional character and his depiction of an all-powerful "Force" is just another Hollywood fantasy, right? Wrong, again, Anakin. Obi-Wan was close to being spot on.

The real "Force" affecting and unifying life and consciousness is the little, itsy-bitsy photon. Photons surround and penetrate us—and the rest of the universe. Indeed, just as Obi-Wan described, photons "bind" the universe together.

In fact, Obi-Wan could have been describing the waves of photons emitted at the time of the Big Bang. Where did this flash of photons go? What happened to them? To answer these questions, we must first understand what light really is. It is not just a candle flame in the darkness.

First, what we call "light" is made up of elementary particles called photons—such as those emitted at the time of the Big Bang. Scientists are only now beginning to answer the following questions:

What are photons? How are they made? What do they do?

The answers to these questions are crucial to understanding how the photonic-flash at the Big Bang could become a source of universal consciousness.

Photons are created in several ways. They are created deep in the core of stars as the internal heat and gravitational force of the star forces atoms to fuse together. This is called fusion. When atoms fuse, the result is a fusion-explosion in which massive amounts of photons are released. These photons are the light created by the sun which gives us both light and life (photosynthesis) here on Earth.

Second, photons are created by biological processes in living tissue. They can be created by biochemical processes in a living body, such as us humans, when changes of energy-states of electrons releases a photon. These human-origin photons create an aura that is detectable with sophisticated scientific instruments. As Masaki Kobayashi of the Tohoku Institute of Technology wrote in his scientific research article's abstract:

> "The human body literally glimmers. The intensity of the light emitted by the body in 1000 times lower than the sensitivity of our naked eyes. Ultra-weak photon emission is known as the energy releases light through the changes in energy metabolism…We found that the human body directly and rhythmically emits light."

Source: Kobayashi M, Kikuchi D, Okamura, H. Imaging of ultraweak spontaneous photon emission from human body displaying diurnal rhythm. PLoS One. 2009 Jul 16; 4(7):e6256. doi:10.1371/journal.pone.0006256.PMID: 19606225; PMCID: PMC2707605.

The photos below were taken by Professor Kobayashi's team using super-sensitive cameras to photograph subjects over twenty-minute periods inside a dark-room. Of these, photos "C" through "G" are the most significant as they show the range of photon emissions off various parts of the subject's head and chest. The red area, found on the head, shows the highest photon emission, up to 3 x 10^3 photons /s/cm^2.

Please note that the calibration bar color-codes distinguishes high intensity photon emission areas of the head from relatively low intensity emission areas. What is important for our purposes is that the human body brain emits and absorbs photons all the time. Being massless, photons pass through our bodies all the time without our ever knowing it.

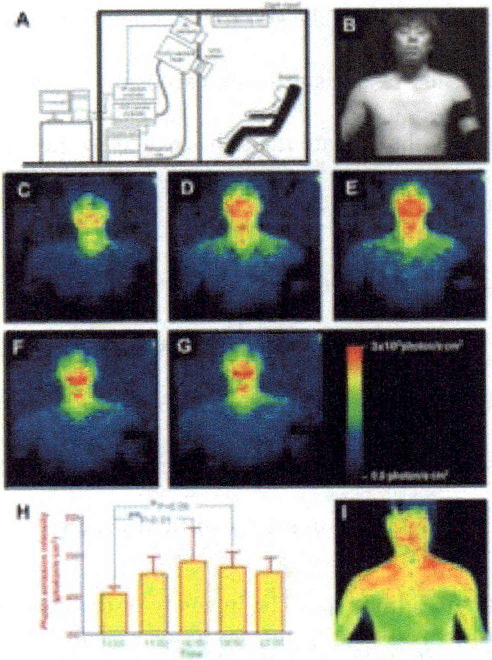

Illustration 2: A. Schematic illustration of experimental setup. B-F. Images of ultraweak photon emission from human body. B. Image of the subject under light examination. C. Image at 10:10. D. Image at 13:10. E. Image at 16:10. F. Image at 19:10. G. Image at 22:10 with a calibration bar which indicates the estimated radiation intensity expressed by photon number per unit of time per unit of skin surface. H. Daily rhythm of photon emission from face and body from 5 volunteers. Significant difference from the photon emission at 10:00am (n=15, Mean plus or minus SD. I. A typical thermographic image of the subject from Fig. 1B-G. Source: Kobayashi M, et al, copyright, 2009.

The human brain is a major emission-source of biophotons. Recent scientific research is unlocking the relationships between brain-emission of biophotons and consciousness. The recent discovery of photonic quantum entanglement has uncovered physiological evidence of what mystics call the "spirit."

"Scientists have an exciting suspicion that our brain's neurons might be able to communicate through light. They suspect that our brain might have optical communication channels, but they have no idea what could be communicated.

Even more exciting, they claim that if there is an optical communication happening, the biophotons our brains produce might be affected by quantum entanglement, meaning there can be a strong link between these biophotons, our consciousness and possibility what many cultures and religions refer to as Spirit.

In a couple of experiments, scientists discovered that rat brains can pass just one biophoton per minute, but human brains could convey more than a billion biophotons per second."

Source: Charles Brass, "Scientists Discover Biophotons in the Brain That Could Hint Our Consciousness is Directly Linked to Light," The Australasian Futures & Foresight Foundation, 2018.

Every photon that passes in and out of our brains carries with it bits of information we process without our ever being conscious of it. This is quantum-information linked directly to the Universal Mind—and, on the quantum mechanical level, with every other life-form in the universe.

This is, I accept, a concept which is difficult to grasp. There is no judgment in anyone not being able to wrap their head around it. Just realize, that whether you can grasp this Universal Mind concept or not, it still exists, and you are part of it.

Photons have different energy levels. The higher the energy level, the higher the frequency and shorter the wavelength. These energy levels are measured by what is known as the electromagnetic spectrum. Below is an illustration showing this spectrum of photon energy levels and how we use each with present technology.

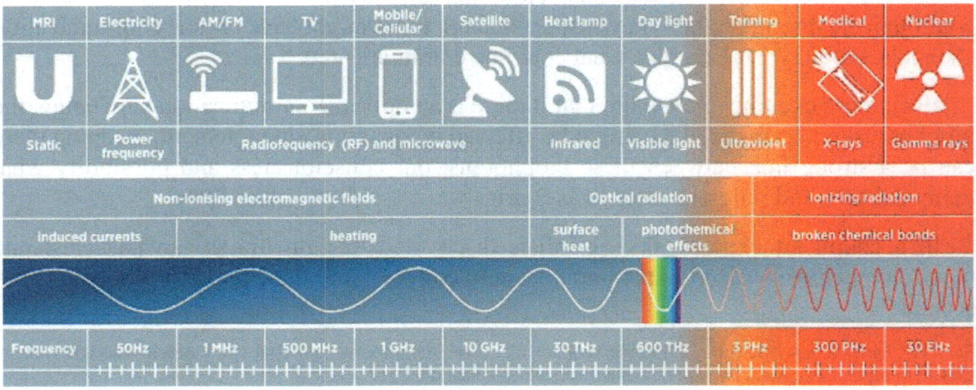

Illustration No. 3. Electromagnetic spectrum. copyright, 2022.

As the above illustration shows, what we know to be "visible" light makes up only a very small portion of the range of energy levels photons can possess. However, these are not necessarily the only energy levels photons may possess.

This spectrum shows only those photon energy states scientists are presently able to detect using present technology. To date, the most energized photon-level detectable is the gamma ray. The lowest detectable photon energy-level is that which we create using strong electric currents and very large magnets for medical-diagnostic procedures and scientific experiments. What photon energy states exist beyond gamma radiation at the high end and static electricity at the lower end of the electromagnetic spectrum are, as yet, unknown.

Gamma rays are the hottest, most energized photons known as yet known to science. Their high energy state releases radiation which can be lethal to humans. Gamma rays are formed from many different types of high-energy events, such as black holes, supernovae, pulsars, fission, fusion, and ordinary radioactive decay.

In addition to their energy levels, Photons possess other properties. First, they move, very quickly. The speed of light varies according to the medium within which the photon travels. Its speed is affected by both refraction and reflection. In mediums such as water or glass, the speed of photons slows slightly. If caught in a black hole, light is trapped and cannot move at all.

In empty space, however, the speed of light (photons) is 186,000 miles or 300,000 kilometers per second. Exponentially, this speed is 3×10^8 meters or 3×10^5 kilometers per second. These measurements of the speed of light will be important to us very soon as we explore the photonic wave's behavior right after the Big Bang.

After the great "decoupling" of photons to matter, the velocity of the subsequent second-flash of photonic energy would have remained at the Hubble Constant of "c" or 180,000 miles/sec. Photons have no "rest energy," as they are forever in motion, unless "captured" by a black hole or absorbed by matter. Photons travel in a straight line, unless reflected or refracted by an intervening medium.

Additionally, photons possess following qualities:

1. They are the most abundant particle in our universe.

2. They possess the dual properties of both particle and wave behavior.

3. Photons do not decay on their own; but they can be absorbed by living tissue (plant or animal).

4. Photons have no mass, and therefore, are only affected by extreme gravitational objects, such as a black hole.

5. Photons are the source of ALL electromagnetic radiation. As photons vary in energy level, the subsequent variation in their wavelength and frequency affects how they interact with the surrounding universe. The electromagnetic spectrum describes and measures these properties. They are "stable" particles, have no electric charge and are -1 spin particles, making them gauge bosons.

6. They carry energy and momentum which are dependent upon frequency of photonic emission.

7. They have interactions with other particles such as electrons.

8. They are created and destroyed by many natural processes, including those occurring within the human brain.

Source: "What exactly is a photon? Definition, properties, facts," Puiu, T., zmescience.com, 2021.

These are not the only properties photons are known to possess; only the easiest properties to detect and measure. As scientists invent more sophisticated instruments, our knowledge of photons and their abilities will astound us all.

For example, we are only now learning about photon "quantum entanglement" and "quantum teleportation." The first serious investigation of these two mysterious photon properties began with Albert Einstein and his research team, in their 1935 research article, "Can Quantum-Mechanical Description of Physical Reality be Considered Complete?" Quantum entanglement is one of the weirder applications of quantum mechanics and creates what Einstein called, "spooky action at a distance."

Source: Einstein, A., Podolsky, B., and Rosen, N., "Can quantum-mechanical description of reality be considered complete," Physics Review, 777, 1935, retrieved at journals.aps.org.

To illustrate what Einstein and his fellow researchers meant, consider the following scenario: two photons interact at the center of the universe and become "entangled" quantum-mechanically. Photon "entanglement" bonds photons quantum-mechanically as though each was part of the other. This is a special characteristic only photons are presently known to possess.

Now, let us assume both photons fly off in separate directions at the speed of light. Time passes and the two photons eventually become so far apart that all contact between each other, in the classical-mechanical realm, is lost. However, a quantum mechanical measurement of either particle will **instantly** provide descriptive information about its partner—no matter how far they are apart—even to the opposite ends of the universe! Instantaneous communication is much, much faster than light speed.

Einstein regarded this capability as "spooky" because it implies that with entangled particles, information can travel instantly between them, regardless of the distance—in a mysterious violation of the natural law that says nothing in the universe (including photons) can travel faster than light (186,000 miles per sec).

Is anything spookier than the thought that photons can communicate instantly between each other from one end of the universe to the other? If light waves can move information that quickly, the idea that "thought" itself is a quantum mechanical property of photonic communication can no longer be ignored. If "thought" exists on a quantum mechanical level, so must consciousness. This leads us to the concept that all our thoughts and our consciousness are a property of photonic radiation. And that this photonic radiation is united through quantum entanglement as part of a greater whole—the Universal Mind.

Imagine all thought, all consciousness being electromagnetically-based. The consequences are staggering, both in their scope and significance to humanity. Just to give some perspective, it means that every living being, (plant as well as animal) everywhere in the universe, is joined together electromagnetically to each other as well as to the Universal Mind, on the quantum mechanical level.

Einstein's 1935 paper was but the beginning of our glimpse into the "spookiness" of the quantum mechanical properties of photons. For example, scientists have been studying "quantum teleportation" for over thirty years. In 2016, a research team reported the successful quantum teleportation of a telecommunication-wavelength photon interacting with another photon over 6.2 kilometers away. The concept of quantum teleportation is complicated, but its origin arises from the logical inferences within quantum-states.

Source: Valivarthi, R. and Puigibert, M. G., et al, "Quantum teleportation across a metropolitan fibre network," Nature Photonics, 676-680, 2016.

Finally, and most importantly for our discussion here, photons carry information. The Pentagon's military scientists appreciate that fact and are busy learning how to take advantage of it for national defense.

"The photon is a fundamental carrier of information, possessing numerous information carrying degrees of freedom including frequency, phase, arrival time, polarization, orbital angular momentum, linear momentum, entanglement, etc. Because optical photons are approximately a million times more costly (i.e., energetic) than their radio frequency counterparts, photons are a valuable resource for many military applications ranging from communication systems to visible and infrared sensing devices."

Source: "Information in a Photon," Defense Advanced Research Projects Agency (DARPA), darpa.mil.com, (no author), (undated).

Not only do photons carry information, but they can also absorb, carry, and transmit information instantly. Researchers at the University of Twente have managed to transmit more than 10 bits of information with a single photon.

Source: Tentrup, Hummel, et al, "Transmitting more than 10 bit with a single photon," Optics Express, Vol. 25, Issue 3, pp. 2826-2833, 2017.

As we have seen from Dr. Kobayashi's research above, the head area of a human subject emits light at the rate of 3×10^3 photons per second per square centimeter. Assuming each photon can carry a minimum of 10 bits of information, we are looking at an incredible amount of information humans absorb and release from photons every second without ever knowing.

This raises a question, however. What happened to the photon-wave emitted at the time of the "decoupling" of photons with matter? These were the same photons which were trapped in the plasma-field during the first 380,000 or so years following the Big Bang. These are the same photons which would have absorbed all the information released as the universal laws and relationships between matter and energy were determined. Where did that light go?

There was no gravitational force to bend light or hold it in place like a black hole. At this stage of the universe, there were no black holes. In fact, at this stage of the universe, the fundamental laws of the universe were only then emerging from the universe's evanescent froth.

But this is not all that was happening. While the universe was determining its own set of fundamental laws, it was also determining the relationships between those laws and all matter and energy that exists or will exist in this universe's future. In other words, the Big Bang was not only a cosmic creation of matter and energy, but an interstellar release of information as well—information that was imprinted upon the photons released at the time of the Big Bang.

So, it wasn't just photons flooding the early universe, but information as well. As the universe formed the laws that were to govern the relationships between matter and energy during these initial 380,000 years; the photons from the initial flash operated as a quantum computer, recording, and measuring it all.

This raises a question. Was the information containing all the natural laws of the universe part of the original photonic-wave when it was released with the Big Bang or was this information released separately and absorbed by the photonic-wave during the first 380,000-year plasma-stage following the Big Bang?

No one knows. It is another of the mysteries surrounding the initial photonic-wave and all photons emitted during that formative stage of the universe.

I favor the theory that information and photonic energy were two separate elements of the Big Bang. The universe is a causal-based organism, its existence the result of cause-and-effect relationships. Unless there was a need for a high energy-high gravity process to take place, the universe would not have experienced the plasma-stage that it did.

The only reason such a plasma-field would have been necessary would have been to bind information to photons and photons to matter. This pressurized "plasma-soup" made the universe a photon-matter coupling system. The photons became "quantum entangled" to information and to the entire universe—creating the Universal Mind.

Each of these initial-flash photons would have absorbed the entirety of all the information being released. Each photon would have absorbed all the information about all that the universe was: all the elements that exist (known and unknown to us); all the elementary particles; and the rules, formulae and laws that would govern each of them.

By determining the laws that would govern its existence and by also determining what matter and energy systems would also exist and how the laws would affect these relationships, the universe determined its own destiny.

Think about it, for a moment. Predicting how a system will behave is simply a function of having enough correct information about the system and its variables. Let's assume I sit at the center of a railroad line. If I determine the length of the track, the number of stations the train must stop at, the number of trains that exist and the speed at which they travel, I can predict with absolute accuracy how fast each train must travel to reach all stations and return within any given time frame.

The algebraic formula for this relationship is: **d = r t** or distance traveled equals rate of travel (speed) times any appropriate time frame. Of course, this mathematical relationship would have been only one of the infinite set of laws and interrelationships determined immediately after the Big Bang.

Let's look at a random sampling of just a few of the universal constants, relationships, and laws that were determined by the Big Bang that would have been absorbed by the photons that were part of that initial celestial flash as information they have carried with them ever since:

Examples of Physical Constants

Speed of light in a vacuum: 186,000 km/sec

Gravitational constant: 6.67408 x 10^{-11} m^3 kg^{-1} s^{-2} ("Precision measurement of the Newtonian gravitational constant," Chao Xue, Jian-Ping Liu, et al, National Science Review, pp1803-1817, Dec. 2020).

Planck's Constant: 6.626196 x 10^{-34} m^2 Js….(a fundamental relationship of quantum physics, multiplying the frequency of a photon by Planck's Constant determines a photon's energy).

Hubble Constant: 69-93 km/sec/megaparsec (expansion rate of the universe).

Atomic mass: 1.660531 x 10^{-27} kg

Electron mass: 9.1093837015 x 10^{-31} kg

Proton mass: 1.67262192369 x 10^{-27} kg

Neutron mass: 1.67492749804 x 10^{-27} kg

Pi: 3.14159265359…(the ratio of a circle's circumference to its diameter)

Phi: 1.618033988749895…(the "Golden Ratio")

Examples of Fundamental Laws of Nature

1. Newtonian laws (classical mechanics)

First Law: An object at rest remains at rest, and an object in motion remains in motion at constant speed and in a straight line unless acted upon by an opposing force.

Second Law: The acceleration of an object depends on the mass of the object and the amount of force applied.

Third Law: Whenever one object exerts a force on another object, the second object exerts an equal and opposite force on the first.

Source: NASA Glenn Research Center.

2. Quantum Mechanics

Just as classical mechanics attempts to explain the universe we perceive with our physical senses, quantum mechanics explores the portion of the universe at the micro-and-submicroscopic level. Most of our present understanding of quantum mechanics comes from scientific experiments attempting to prove or disprove theories of quantum mechanical interrelationships developed by Maxwell, Einstein, Bohr, Heisenberg, Feynman, Schrodinger, Planck, de Broglie, Dirac, Pauli, Susskind, Hawking and others.

In 1785, Antoine Lavoisier discovered the Conservation Law of Mass, which states that matter can neither be created nor destroyed. In 1842, Robert Mayor discovered the Conservation Law of Energy, which is now at the core of the First Law of Thermodynamics. The Conservation Law of Energy states that energy can neither be created nor destroyed.

In the twentieth century, Albert Einstein's paper, "Relativity: The Special and General Theory" (published in 1920 by Henry Holt and Co.) became the creative spark behind what became known as quantum mechanics. Einstein's discovery that $E = mc^2$ (energy (E) equals mass times the square of the speed of light) meant that energy and mass are equivalent; their total remains constant forever. They are interchangeable depending upon the interactions between them, but the total amount of mass and energy remains the same. There will be no more. There will be no less. What mass and energy exists now, has existed since the universe began and will remain in existence until the universe ends.

3. Thermodynamic Laws

(1) First Law of Thermodynamics (new)

"All real-world processes involve transformations of energy, and the total amount of energy is always conserved (is neither created nor destroyed). The universe is a continuum of energy which remains the same regardless of transformation. As to energy, time is symmetrical. The same energy exists for all time, simply changing states over time."

(2) Second Law of Thermodynamics (new)

"In all natural processes, the total entropy of a system plus its environment cannot decrease. It can remain constant for a reversible process but must always increase for an irreversible one. The universe acts spontaneously to minimize potential outcomes of any process. Therefore, the universe always seeks to maximize disorder (entropy). As to entropy, time is asymmetrical. Therefore, the Second Law of Thermodynamics creates a measure of the transformation of energy over time."

Source: "All about entropy, the laws of thermodynamics and order from disorder," Archives of Science, 2001, reprinted from entropylaw.com.

Originally, the word "entropy" meant randomness or uncertainty of physical systems such as the universe. This meant that as time passes, the universe becomes ever more random as energy and matter continue to expand away from each other.

However, in 1948, American mathematician C.E. Shannon published "A Mathematical Theory of Communication." This paper forced scientists to view uncertainty far deeper than ever before. Uncertainty was caused by change and change implied an increase in a heretofore neglected causal relationship that would turn the Second Law of Thermodynamics—and physics—on their heads.

This causal relationship was simply defined as:

Any change in the state of a closed-system increases its information. Any increase in information increases the system's uncertainty or entropy. Information, therefore, equals entropy. Shannon's paper proved thermodynamic laws applied to information sources and that information communication could be expressed mathematically.

Source: "A Mathematical Theory of Communication," Shannon, C.E., reprinted from The Bell System Technical Journal, Vol. 27, pp. 379-423, 623-656, July, October, 1948.

Shannon's redefinition of closed-system entropy to include information (the measurement of changes of states of any system over time) rivals Newton's observation of gravity in the fall of an apple as one of the greatest intuitive leaps of intellect in human history. His new approach changed the way we apply information to our existence in the universe. The logical inferences are startling.

Dr. Leonard Susskind, Ph.D., the noted physicist, Stanford University physics lecturer, and author believes the Conservation Law of Information is "undoubtedly the most fundamental of all physical laws.

"The conservation of information is simply the rule that every state (of existence) has one arrow in and one arrow out. It ensures that you never lose track of where you started."

Source: Susskind L & Hrabovsky G, THE THEORETICAL MINIMUM: What You Need to Know to Start Doing Physics, pp 9-10, Basic Books, NY, 2014.

To take Dr. Susskind's analysis one step further, the Conservation Law of Information ensures that every bit of information released by the early universe can never be lost. Susskind's "arrow of time" connects the universe's beginning with the present and into the infinite future, as well.

If, at the Big Bang, the universe contained all the information that would ever exist, then it must have also possessed all the probable futures that the applications of that information would create—including all matter and energy forms. This necessarily includes all life forms in the universe—including human life.

Therefore, the photonic-wave would have become a "living" quantum-mind as large as the universe, capable of predetermining the existence of every form of life and predict its future as well.

Imagine all photons of the photonic-wave creating a "cloud" of photon particles covering the entire universe. As photons are capable of "quantum entanglement," every photon in the universe is instantly aware of the status of every other photon it is entangled with. This means every photon is capable of spontaneously knowing everything that happens anywhere in the universe because it exists everywhere and is interconnected everywhere.

But that is not all. The photonic-wave encapsulates the entire universe, and it exists in a quantum state, capable of spontaneous data transfer throughout the universe. Therefore, every living being within the universe is also interconnected—at an ultra-long wavelength and ultra-low frequency which are, as yet, undetected by present technology. It's most likely there, but beyond the scope of our present technology.

This signal may be hidden within the Cosmic Microwave Background radiation itself. Perhaps the CMB acts as a carrier-wave and the Universal Mind's signal is hidden within it. The CMB's heavy static tends to support that conclusion. Our instruments are presently inadequate for the more fine-tuned adjustments necessary to detect it, however.

CHAPTER 7

A Little "Light" Reading

I know these concepts are not easy. But we are taking this step-by step. So, hang in there with me. Let's recap where we have come before we move on.

The Big Bang produced an immense flash of photons and matter in the form of subatomic particles. For the first 380,000 years, these photons were trapped within a plasma-field "universal womb". These trapped photons absorbed the information released as the universal laws and matter-energy relationships were determined. The photons spread out within this "womb" like a cloud or membrane of living tissue.

Two significant events occurred during the "womb" phase of the universe. First, the photons in the cloud-membrane became "quantum-entangled." The proximity of photons to each other "entangled" them quantum mechanically.

This "entanglement" resulted in each photon sharing the information it had absorbed with every other photon. Every photon within the "universal womb" shared all the universal laws and relationships that had been created within that "womb."

Second, this "quantum-entanglement" resulted in the photon cloud-membrane becoming self-aware. The universe-wide sea of photons shared both information and self-awareness. This sea of linked photons became the "Universal Mind" either during the initial de-coupling era or shortly after the photons were released as the de-coupling era ended.

Then, after approximately 380,000 years, this plasma-field "universal womb" cooled to the point where photons were free to move at their present speed—186,000 miles a second. This second burst of photonic energy filled the universe with the photons from the "universal womb." As the universe expanded over the next 13-plus billion years, the Universal Mind and its photon-cloud expanded as well, filling the present universe with consciousness and self-awareness.

That is the theory. Is it valid?

The answer to that question depends upon that tiny, massless ionized particle—the photon. What can it do? What properties and capabilities does it possess? How much do we know about those first photons from the early universe?

Assume there are three sets of photons at work in the universe:

A. The original photonic-wave that bathed the universe in information from the Big Bang;
B. Photons created by stars, our sun and other interstellar phenomena on a daily basis; and
C. Photons created by localized life-forms (such as humans), weather (lightning) and man-made electronic sources such as electric motors, etc.

Our technology can measure the speed and behavioral characteristics of "B" and "C" photons, but what about the "A" photons, the ones from the Big Bang itself? They are the mystery.

All we know about the original photons is what COBE, WMAP and Planck have shown us which is very little. The universe has been bathed in these "A" type photons for billions of years. Yet all we know about these ancient photons are the images and data dating back 380,000 years after the Big Bang—and nothing since. That is like trying to estimate the atmospheric pressure at the bottom of the ocean by sticking a toe into the surf on the beach.

There is no evidence to suggest the speed of light versus the speed of the expanding universe changed in the 13-plus billion years that has followed the second photon-flash. Carrying the math forward in time 13 billion years, we find that the photonic-wave must have traveled very far indeed—approximately 1.2299×10^{29} km.

Keep in mind that the expansion rate of the universe itself is between 69-92 km per second per megaparsec. Photons travel much faster; 300,000 km per second.

While the photons travel much faster than the universe expands, the photons cannot escape the universe. The photons are contained by the lithium-metallic skin surrounding the universe. This lithium metallic skin acts as a barrier. Photons do not penetrate the lithium. They are either absorbed by the lithium or reflected back toward the universe center.

The universe has been bathed by this information-rich photonic "ocean" for 13 billion years. The universe is literally saturated by the information released from the Big Bang itself.

Every cell in our bodies is made of matter and energy created and then saturated in this ocean of information-rich photons from 13 billion years ago. Further, this sea of photon-encoded information exists all around us, even today.

We cannot "see" this information with our human eyes; or "feel" it with our human touch. This information permeates the present universe—but on a level our senses cannot reach—the quantum mechanical level.

One of the issues we need to address is the lack of appreciation most people have for how ubiquitous photons are to our daily lives. Most people are aware that sunlight is made up of photons. But that's not all photons are about.

Every day of our lives, we are literally bathing in an ocean of photons. Some of this ocean of light is comprised of "new" photons, just emerging from our Sun. Other photons in this ocean of light are old—some so old they originated at the time of the Big Bang, 13 billion years ago.

Photons vary in their energy level, which we detect by measuring their wavelengths and frequencies. Sunlight, the visual part of the light spectrum, is a but a small part of the photonic impact upon life. Being massless, photons pass through our flesh without our ever feeling it. We can feel it's thermal properties as warmth on our skin. But the effects of photonic energies go much deeper.

As described in his treatise, *Interaction of Radiation with Matter*, Dr. Perry Sprawls, Ph.D. explains that:

1. Photons can penetrate living tissue without interacting with it. They can just pass right on through our bodies without leaving anything behind.

2. Photons can also penetrate tissue and be absorbed by the body (including the brain), also without our detecting it is happening.

3. Lastly, photons can penetrate tissue and then become scattered waves of radiation within the body, as it is deflected from its original direction, and leave part of its energy behind in the tissue as it passes through. This scattering of radiation can change the structure of our DNA or cause normal cells to turn cancerous.

Source: Sprawls, Dr. Perry, *Interaction of Radiation with Matter*, www.hep.ph.ic.ac.uk.

We have already discussed the most energetic (and most dangerous) photons, the gamma rays. As the photonic wavelengths increase and the frequency deceases, photons are emitted as x-rays. These x-ray photons, while harmful if too many are absorbed, are used routinely in medical procedures such as x-ray imaging and CAT scans. However, both x-ray and CAT scans can affect human DNA and cause cancers in some patients due to the radiation scattering effect just discussed.

Slowing the frequency further still lengthens the photonic wavelength to the ultraviolet range of light. Ultraviolet wavelength photons can cause sunburn.

As photonic wavelength increases and frequency correspondingly decreases even more, photons emerge (in order) as visible light, infrared, and radio waves.

Radio wave photons carry radio, TV, FM, AM radio signals, and the Internet. More to our point, scientists have learned that photons can carry data. Optical Data Transmission (ODT) has replaced radio waves or electrical wires with photons.

"Data travels over the light beam using a modulation method known as frequency-shift keying or FSK. FSK rapidly switches the IR (infrared) light between two different modulating frequencies. One frequency represents a logic 1 value while the other frequency represents a logic 0 value...Optical data couplers use a beam of infrared light to transmit digital information between moving objects without the problems associated with wired or wireless systems."

Source: Sensor Sense: "Optical data transmission," edited by Robert Repas, MachineDesign, 2007.

Fiber optics carries our digital data from smart phones to Internet.

"Fiber optic Internet is an Internet connection that transfers data fully or partially via optic cables. 'Fiber' refers to the thin glass wires inside the larger protective cable. 'Optic' refers to the way the type of data is transferred—light signals." Data is transferred almost at the speed of light, as some speed is lost because the signals bounce around inside the glass wires.

Source: Fiber "Optic vs. DSL Internet: How they work plus pros and cons," attinternetservice.com and by broadband.gov. (undated).

All these signals the photons are carrying, in the paragraphs above, are information-loaded. If photons can carry radio and television programming, including audio and video, photons can just as easily communicate—or share—consciousness.

CHAPTER 8

Mirror in My Mind

There is a potential problem, however. We'll call this potential problem the "two-state paradox." Let's understand it first, and then see if there is a solution.

The "two-state paradox" accepts that any living organism is best suited for the environment in which it first came into existence. Dinosaurs evolved in a very specific environment, as did mammals. Change the environment and the quality of existence of a species declines. Change the environment of a species drastically enough and its extinction becomes inevitable.

This rule applies to all living things because every living species is a biological system which evolved from very specific adaptations caused by very specific environmental conditions. If water seeks its own level, so too do species—and self-aware photonic-fields.

A photonic-field which became self-aware as the result of a particular physical environment would not adapt well to any significant change in that environment—unless adaptations came into play to preserve the necessary balance.

During the initial plasma-era of the universe, the photon-field became self-aware. But that plasma-field created its own unique state of existence within which the photonic-field was able to absorb information and, through quantum-entanglement, become self-aware.

The conditions present at the time the photon-field became self-aware are known. Photon-field self-awareness occurred while the universe was caught in this plasma-era with super-hot temperatures and intense gravitational forces forcing all subatomic particles, including photons, to mix in a plasma-soup. The region consisted of high-density, highly-energetic, charged particles crackling with energy levels approaching infinity.

Wouldn't this state of self-awareness be dependent upon the existence-state in which it was created? That existence-state would include the extreme temperature and gravitational gradients which existed within the plasma. What would happen to the self-aware, photonic-field once this temperature-gravitational state changed allowing photons to move freely?

Did the Universal Mind survive the second photonic-wave flash? The answer depends upon whether the Universal Mind could adapt from existing in a state it preferred—high density, high temperature and gravitational forces—to an existence state of relatively lower density, temperature, and gravitational forces.

As the universe and the self-aware photon-field expanded together, did the universe evolve and adapt so that the quantum-entangled photons carrying the Universal Mind could maintain coherence? Could the universe evolve to allow the Universal Mind to adapt to the new environment created by the expanding universe?

It appears that the universe did just that. Recent observations suggest that the universe created a living brain structure within which the self-aware photonic-field has been able to retain its coherence despite the thirteen-billion-plus years of interstellar universal expansion.

How did this happen?

Professor Franco Vazza (astrophysicist from the University of Bologna) and neurosurgeon Dr. Alberto Feletti (University of Verona), recently investigated to what extent human brains mimicked the cosmic web—that vast expanse of interstellar space and all its matter (megaparsec long filaments of galactic clusters, gas clouds, voids, and dark matter) which is the universe. What they found was an astounding similarity between the structures of each.

For example, they discovered that the human brain's complex temporal and spatial structure could be modeled as a hierarchical network in which "neurons cluster into circuits, columns and different interconnected functional areas." This structure within the neural network allows the linking between different areas of the brain to coordinate and integrate data-flow between basic life functions, perception of environment, emotion, judgement, behavioral response, and memory.

Vazza and Feletti discovered that this human brain structure mirrors the way the universe is organized:

> "The most important building blocks of the cosmic web are self-gravitating dark matter dominated halos, in which ordinary matter has collapsed to form galaxies (and all the stars with them)…and has developed into larger groups or clusters of galaxies, filaments, matter sheets and voids, in a large-scale web in all directions in space."

Vazza F. and Feletti A., "The Quantitative Comparison between the Neuronal Network and the Cosmic Web", Frontiers of Physics, (2020).

The Vazza-Feletti team compared microscopic images of brain tissue with computer-generated, simulated "slices" of the observable universe and compared the two.

The "observable universe" refers to the maximum distance into the universe which we can detect photon radiation (light) either by radio or visual telescopes. In visible light, the further backward in time we can observe is almost to the Big Bang itself—13-plus billion years in the past.

Source: NASA at science.nasa.org. (no date)

What they discovered was an amazing similarity between the universe and the human brain. Consider the following astounding similarities:

1. "Cortical gray matter (representing over 80 percent of brain mass) contains about 6 billion neurons (19 percent of brain neurons) and nearly 9 billion non-neuronal cells. The cerebellum has about 69 billion neurons (80.2 percent of brain neurons) and about 16 billion non-neuronal cells. Interestingly enough, the total number of neurons in the human brain falls in the same ballpark of the number of galaxies in the observable universe."

2. Approximately 2.6×10^{12} (2.6 trillion) galaxies may be present within the observable universe with up to 5×10^{10} (50 billion) galaxies with masses equal or larger to our Milky Way galaxy. Therefore, there is a close comparison in the number of human brain neurons to cosmic galaxies.

3. Neural brain tissue mimics the long filaments of ordinary and dark matter that extend several tens of Megaparsecs to connect clusters and groups of galaxies.

4. The human brain and the universe are organized into discrete networks and both galaxies and the brain's neural network have a typical scale radius, which is only a very small fraction of the typical length of filaments they are connected to. At the highest magnification, the human brain's neural network and connecting filament tissues start presenting some similarity with the cosmic web on a scale of greater than 20 megaparsecs.

5. Comparing a cosmic web slice with a human brain tissue slice, the Vazza-Feletti team found a remarkable similarity with spectra obtained when comparing 1mm or smaller scales in brain tissue to 100 or smaller megaparsec scales of the cosmic web.

6. In both the human brain and the universe, information flow and energy exchange account for less than 25% of the total mass-energy content of each.

Source: Vazza F. and Feletti A., "The Strange Similarity of Neuron and Galaxy Networks," Nautilus, 2017.

"Strikingly, in both cases (human brain and universe) approximately 75% of the mass/energy distribution is made of an apparently passive material, that permeates both systems and has an only indirect role in their internal structure: water in the case of the brain, and dark energy in cosmology, which to a large extent does not affect the internal dynamics of cosmic structures."

Source: Vazza-Feletti, Frontiers (cited above).

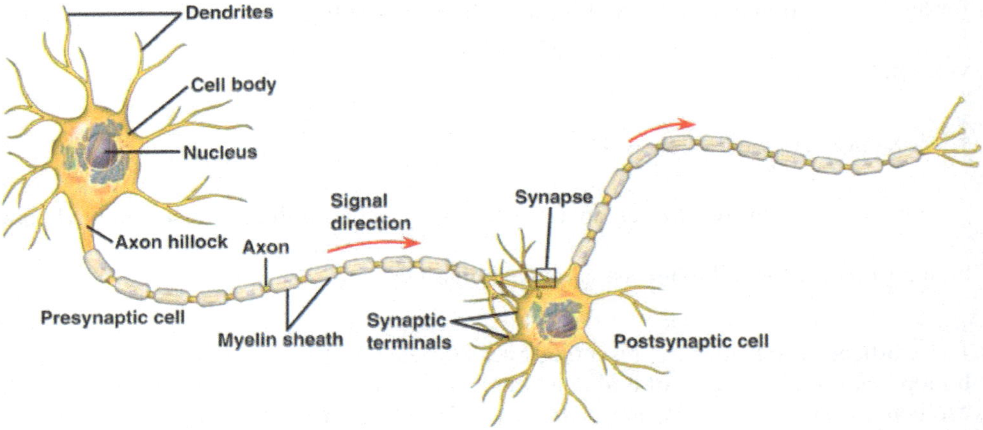

Illustration No. 4. Artist's drawing of neurons. copyright 2022.

The above illustration is an artist's rendering of the typical neuron found in the human brain. The neuron consists of a nucleus, cell body (soma), dendrites, axon covered in its myelin sheath and axon terminal buttons connecting to axons from other neurons.

other neural tissue and to what actual neurons look like in a tissue slice taken from a brain. The tissue slice has been color-stained as an observational aid.

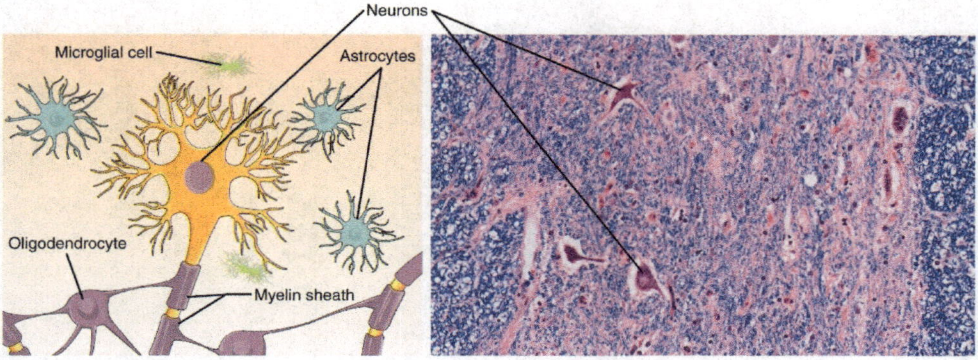

Illustration No. 5. Artist's drawing of a neuron and actual photo of neural network in a slice of brain tissue. Copyright 2022.

In Illustration No. 5, above, we transition from an artist's drawing of a neuron among The next image is from Vazza-Feletti's article, "The Strange Similarity of Neuron and Galaxy Networks," published in *Nautilus* magazine in 2017 (cited above) .

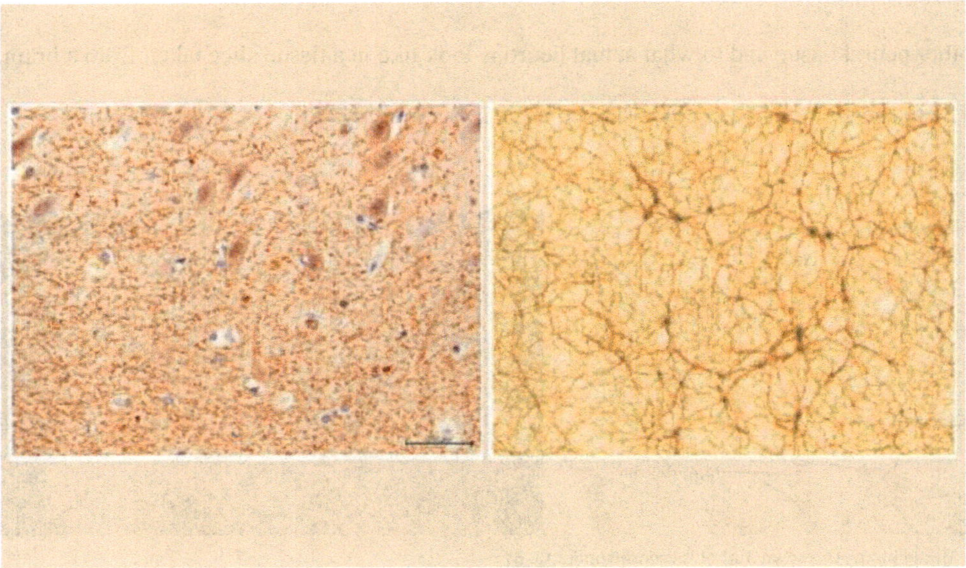

Illustration No. 6. Photo of neural tissue (left) compared with cosmic web (right). The slice on the left is a microscopic image of a slice of neural brain tissue from the cerebellum. The neural bodies have been color-stained with clone 2F11 monoclonal antibody against neurofilaments. The image on the right is a computer simulated matter distribution of the cosmic web. Photo by Automated Immunostainer Benchmark Xt, Ventana Medical System, Tucson, Az. copyright 2017.

The Vazza-Feletti team attempted to make each of the above-shown slices proportionate. The image on the left is a of a four micrometer-thick slice of brain neural tissue from the cerebellum. On the left, is an image generated from a computer-simulated slice of the cosmic web one billion light years across.

But the neurons in the above illustrations are shown embedded inside gray matter of the brain. How would the brain's neural network appear to someone small enough to be on the inside of the brain and with the gray matter removed?

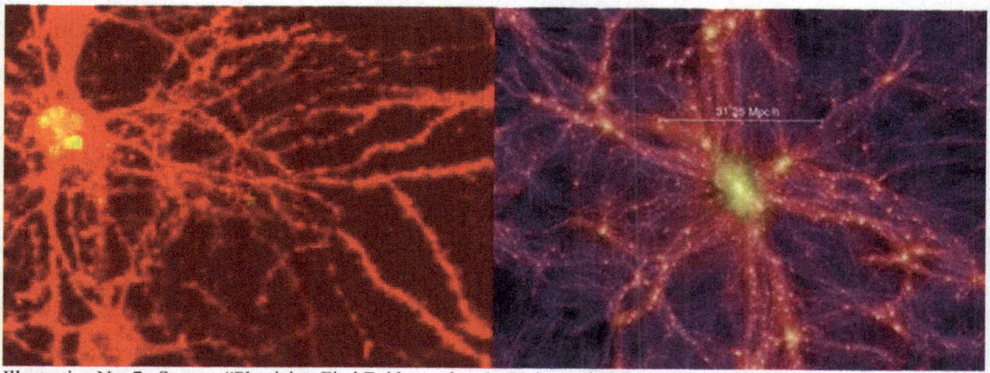

Illustration No. 7. Source: "Physicists Find Evidence that the Universe is Like a Giant Brain," (no author), published by anomalien.com, copyright 2020.

Examine the above illustration. Can you tell which image is from brain tissue and which is the universe? The image on the left is from a mouse brain, color-stained to eliminate the gray matter and greatly enlarged. Three neurons can be seen on the left side of the left image, two others are stained red on both the top and bottom. The third neuron is on the middle-left, shown in yellow.

This image provides a view as though one were inside the brain with the gray matter removed. With the gray matter removed, the similarity between the neurons of the brain and the cosmic web (image on the right) is much more apparent. The image on the right is of a large galactic cluster (yellow center) surrounded by stars, individual galaxies, and dark matter. Source: anomalien.com (cited above).

Now, let's reconsider the "two-state" paradox that we discussed briefly at the beginning of this chapter. How could the photonic-field which became the Universal Mind, retain its self-awareness or coherence as photons became freed from the plasma and expanded with the rest of the universe? Put another way, as the universe evolved and expanded over time, how was the Universal Mind able to adapt and remain self-aware?

The Vazza-Feletti team unknowingly provides the answer. The answer lies in the structure the universe evolved into as it expanded. Structure is an environment. An environment becomes a state of existence. As we discussed at the beginning of the chapter, the environment a species is born into is its best hope for survival—unless the species can adapt as the environment changes states.

That is exactly what happened with the Universal Mind as its environment changed from a state of high-density, high-energy, high-temperature, and high-gravitational force to a state of gradual decline of all four of these variables as entropy increased over time. The structure of the universe did not occur quickly, but over billions of years. This gradual change of state gave allowed the Universal Mind to adapt.

We need look no further than the stunning images comparing the structure of the human brain to the cosmic web to realize the Universal Mind was, indeed, successful at adapting to a new state-of-existence. The proof that this effort was successful is evidenced in how closely the structure of the human brain is to its evolutionary parent—the Universal Mind.

Allowing for differences in scale, how closely the human brain and the cosmic web compare with each other in terms of structure cannot be a mere coincidence. Now that we humans have finally managed to build instruments sophisticated enough, we can at long last see the Universal Mind staring back at us in the cosmic mirror.

In this chapter, we have examined how the universe and the human brain mirror each other's structure. Galactic clusters perform the same functions as neurons. Web-like strings of gas-cloud filaments perform the same function as connective tissue and neuralfilament within the brain. Neuroscientists have recently recognized the importance of neurofilaments in synapses and their relationship to impaired brain function due to injury and disease.

"Neurofilaments are major cytoskeletal components in mature neurons. They are found in the cytoplasm of neurons within the peripheral nervous system and the central nervous system, most abundantly in axons, but also in cell bodies, dendrites and synapses. They are expressed more highly in large, myelinated axons, **where they are organized in parallel structures maintained by side arms projecting outwards from a filament core.**" (emphasis added).

Source: "Neurofilaments: neurobiological foundations for biomarker applications," Gafson A, Barthelemy N, et al, Brain, published by Oxford University Press, July, 2020.

Please give special attention to the above-description of "neuralfilaments" found in the human brain. They possess parallel structures with side arms projecting outwards from a filament core. Does this not sound very much like the long-strands of gas-cloud filaments we saw in the computer-simulations of the cosmic web?

Both the cosmic web and the human brain possess similar structures. Neurons mimic galactic clusters. Strands of gas-cloud filaments, megaparsecs long, remind us of the neuralfilaments connecting neural nodes in the brain. And ever-present among both the cosmic web and the human brain are the photons—particles we take so much for granted, yet which hold the key to our consciousness and individuality.

Without information to store, memory is of no relevance. Without memory, there can be no unique information within each person. Without possessing unique information, we do not possess individuality. Without individuality, there is not unique information to conserve and the Conservation Laws of Information and Energy do not apply to us.

The universe created its own brain structure. This was no coincidence. A brain structure which evolution copied when it created the human brain. This was no coincidence, either. Both took foreknowledge of what was necessary for the Universal Mind to adapt to changes within the universe and for intelligent life to evolve on Earth—as it must throughout the cosmos.

How can this happen? What intelligence is behind the creation of the cosmic web's brain structure? I will give you a hint: being reborn is not just for species living within the universe. The universe itself has been reborn many times.

CHAPTER 9

The Reincarnation Machine

In the previous chapter, we examined some of the ways human physiological evolution mimics the structure of the universe. This should not be a surprise. The universe has made us in its image. The universe has had over 13 billion years to evolve its own structures, is it any wonder it has shaped our own?

In the previous chapter, we also discussed "states-of-being" and how species survival depends in large part upon how it adapts to changes in its environment. We also brought up the subject of structures and how structures can be used to help a species survive—or a Universal Mind to retain its coherence.

Reincarnation is just such a structure. It has evolved over time as an adaptation to conserve our unique individuality—created by the information and self-awareness we share with the photonic-wave that has become the Universal Mind. If Reincarnation exists as part of the human experience, it must have originated from the Universal Mind from its own experience. In other words, if Reincarnation evolved in humans, it must be a core-function of the universe. It must be a fundamental consequence of natural law.

Why must this be?

Natural law controls what occurs within the universe. We know the universe is just another system. It is vastly bigger, and vastly older than any other system; but it is still just another system. All systems follow natural laws. We don't understand all of these laws yet, but we are steadily learning more and more about how the universe works.

If Reincarnation is part of the human experience, it must also be a part of the universe's experience as well. Is there any evidence this is so? Is there any evidence the universe has itself been reincarnated or been "born again?"

There are several theories being investigated of how our universe has been reborn many times in the past and will be reborn in the future.

One theory on how our universe will end is called "The Big Crunch." The universe begins with a "Big Bang," expands until gravitational forces overcome the force of the expansion. The universe then contracts until it reaches its ultimate end—a black hole singularity. This singularity then creates the next Big Bang and a new universe is born.

Source: William Harris, "How the Big Crunch Theory Works," 2 March, 2009, republished in HowStuff Works.com, 21 October, 2021.

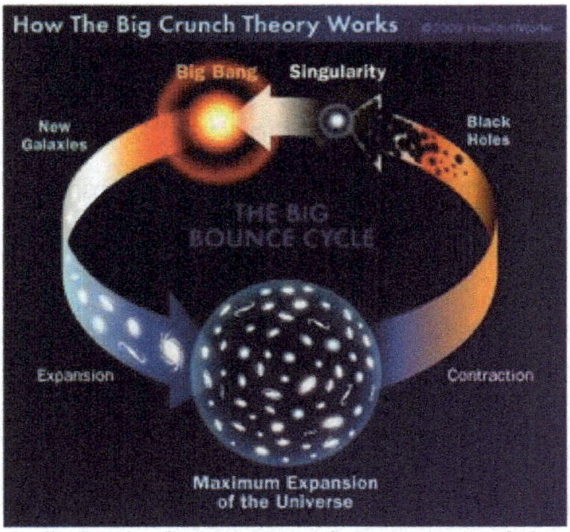

Credit: Creative Commons at creativecommons.com & Quora at quora.com.

Illustration No. 8. The Big Crunch. copyright 2022

According to British mathematician, astrophysicist, quantum physicist, and philosopher Sir Roger Penrose, the universe has been reincarnated an unknown number of times. Since 1973, Penrose has held the distinguished title of the Rouse-Ball Chair of Mathematics at Cambridge University. Knighted for his services to British science in 1994, Penrose was awarded the Nobel Prize in physics in 2020 for his work in black hole cosmology.

Penrose became interested in human consciousness, as a "state-of-being" and wrote two books in which he argued that the reality of consciousness required a quantum mechanical formula: *The Emperor's New Mind* (1989) and *Shadows of the Mind* (1994).

In his 2010 book, *Cycles in Time: An Extraordinary New View of the Universe*, Penrose argues that the universe has experienced repeated cycles of birth, life, death, and rebirth. He believes that the universe is a systemic structure in which the death of the previous universe becomes the Big Bang of the next. This theory, which he has termed, Conformal Cyclic Cosmology or "CCC," proposes that the universe exists in a "state-of-being" of perpetual reincarnations.

How does Penrose explain the transition from the expansion of universe "A" to its "death" and rebirth as successor universe "B?" Penrose did not originally believe in a universal contraction (a "Big Crunch"). Instead of a Big Crunch, Penrose's alternative appears to be that an expanding universe reaches a point where its internal pressure becomes inconsistent with further expansion and creates the next Big Bang from the over-pressure against the fabric of space-time within the older universe. A rip in space-time.

Source: R. Penrose, "The Basic Ideas of Conformal Cyclic Cosmology," AIP Conference Proceedings, Frontiers of Fundamental Physics, American Institute of Physics, 2012.

Let's see if we can visualize this. Consider a string of balloons. Each balloon is connected to the one behind and the one in front. Inflate the first balloon and it expands until it reaches a critical point where the balloon fabric is no longer strong enough to hold the air inside without bursting.

Then the balloon deflates explosively. The force of that explosion pushes tears the skin of the next balloon. The pressure from the deflating first balloon forces air from the first balloon into the second balloon. This process can continue as long as there are balloons connected to the one which bursts or the balloon from the original universe is somehow introduced into the next universe.

The model which most fits what occurred during our universe's Big Bang is the latter. What we have in our universe is a very thin, very light lithium skin that emerged out of the Big Bang with the helium and hydrogen.

That our universe began with a Big Bang, out of which only a few elements emerged (hydrogen, helium and lithium) plus photons, suggests that on the other side of that Big Bang would have to be a Big Crunch strong enough to reduce all matter into those three basic elements. Keep in mind that, in this model, the Big Bang is nothing more than a rip in space-time at the point where the lithium skin of the dying universe tears from over-pressure.

The over-pressure from the dying universe is much greater than the vacuum existing in the empty-fetal universe. This pressure differential creates the force by which the contents of the dying universe are thrust into the new one. All natural laws of the dying universe are carried into the new universe as well. This cycle can go on indefinitely as all matter, energy and information remain conserved.

However, as seen in the illustration below, Penrose has since accepted a form of the Big Crunch in his cyclic cosmological model.

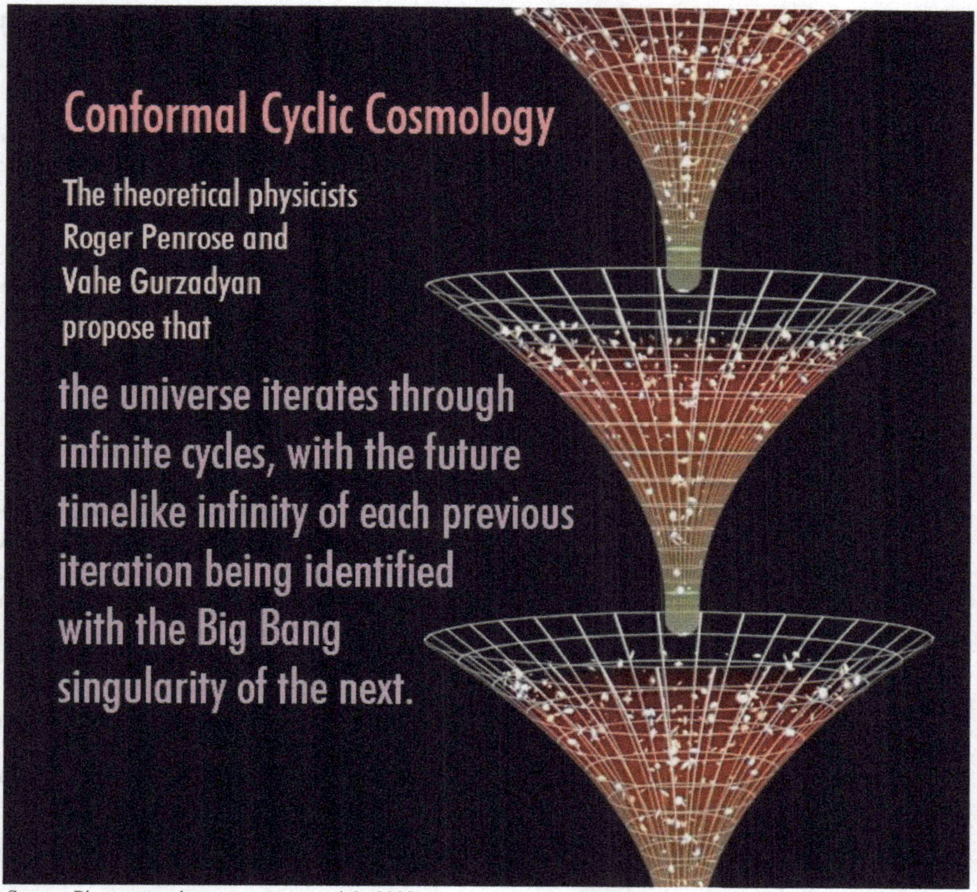

Source: Pinterest at pinterest.com. copyright 2022.

Illustration No. 9 Prof. Penrose's Conformal Cyclic Cosmology. Professor Penrose's Conformal Cyclical Cosmological theory as a Big Crunch. Each funnel depicts a universe at maximum expansion. What follows is a gradual contraction to the point of singularity, which spawns a new universe. Theoretically, this process has the potential for infinite rebirths because all the information, energy and matter from the old universe is conserved and recycled into the next universe.

A third theory of universal rebirth has been proposed by Dr. Paul J. Steinhardt, the Albert Einstein Professor in Science at Princeton University and Dr. Anna Ijjas, of the Institute for Theory and Computation, Harvard-Smithsonian Center for Astrophysics. Their theory has been called the Big Bounce theory of cyclical universes or "Bouncing Cosmology."

The Steinhardt-Ijjas theory assumes that universes contain an energy beyond that which we can observe at present. This energy is manifested as a cyclical oscillation within the fabric of the universe which repeats between regular phases of expansion and contraction (thus the "Bounce"). Again, all natural laws, matter, energy and information are conserved from the older to the newer universe.

Their theory also considers the impact of pressure differential as a major contributor toward both universe expansion and contraction:

"Yet another generic prediction is that the current vacuum and accelerated expansion are temporary. That is, the vacuum is either unstable and slowly changing with time; or metastable, protected by an energy barrier that will eventually be bypassed through quantum tunneling. The possibility that we live in an unstable or metastable phase is well-motivated….The current accelerated expansion phase (of the universe) cannot be eternal. The universe must eventually transit to the true vacuum and begin to contract. Will that bring the universe to a crunch (an end) or a bounce? We have shown a bounce back to an unstable or metastable vacuum with positive energy density is possible."

A. Ijjas and P. Steinhardt, "A new kind of cyclic universe," pub. online at arxiv.org 2019.

Each of these three theories of cyclical universe rebirth is consistent with the human process of Reincarnation described in detail in my previous book, ***REINCARNATION: The Scientific Theory***—but on a universal scale.

CHAPTER 10

Dancing with Siddhartha

For the approximately 500 million Buddhists around the world (250 million in China alone), however, Reincarnation and universal consciousness are not mere theories. They are both accepted as fact, although Buddhists consider the term "reincarnation" to be a Western bastardization of the Buddhist concept of "rebirth."

According to Buddhist teaching, the concept of "rebirth" is not limited to individuals. As Tenzin Gyatso, the XIV reincarnated Dalai Lama and spiritual leader of all Buddhists, explains:

"The infinite process of evolution is very similar to the modern scientific notion of the Big Bang. If the scientific cosmological theory of the Big Bang accepts only one Big Bang as the beginning of everything, then of course this would not fit in with basic Buddhist cosmology…However, if the (scientific) Big Bang theory does not entail only one Big Bang at the beginning, but accepts a multiplicity of Big Bangs, then that would correspond very well to the Buddhist understanding of the evolutionary process."

Source: Dalai Lama, A Simple Path, Thorsons, Hong Kong, p.44, 2003.

A "multiplicity of Big Bangs"? Yes, indeed. Buddhist doctrine accepts the existence of a series of universes; each one a rebirth of the one preceding it, infinitely. This is one of the core concepts of Buddhist cosmology and is consistent with the companion concept of rebirth of the individual.

Does this fundamental Buddhist concept remind you of anything? Doesn't it remind you of Professor Penrose's concept of universal "rebirth" termed "Conformal Cyclic

Cosmology" and other theories of cyclical universe rebirth? Does this not remind you of the theory of universal Big Bang followed by what cosmologists call the Big Crunch where our present universe is part of an endless series of universes and where the disintegration of an old universe becomes the Big Bang of its successor.

Source: R. Penrose, Cycles of Time: An Extraordinary New View of the Universe, Knoph, New York, 2011.

Sir Penrose's theory is based upon both logic and mathematics. It has earned him world-wide acclaim. His perception of the universe is very consistent with traditional Buddhist teaching. As he explains in the BBC television documentary on Stephen Hawking's book, *A Brief History of Time*:

"There is a certain sense in which I would say the universe has a purpose. It's not there somehow by chance. Some people take the view that the universe is simply there and it runs along—it's a bit as though it just sort of computes and we happen by accident to find ourselves in this thing.

I don't think that's a very fruitful or helpful way of looking at the universe. I think that there is something much deeper about it, about its existence, which we have very little inkling of at the moment."

Source: Interview of R. Penrose, "A Brief History of Time," Anglia Television Channel Four Films, distributed by Triton Pictures, 1991.

To understand and conceptualize the Universal Mind, it is necessary to consider Buddhist teachings. This is not an endorsement of Buddhism. However, studying some basic Buddhist concepts will help us understand how the Universal Mind and the human mind are quantum entangled.

Many of the terms used in Buddhism are Tibetan and therefore somewhat alien to Western culture, thus making them difficult for us to follow. However, the fundamental principles, when broken down to their roots, provide a commonsense approach to understanding the relationship between the Universal Mind and Reincarnation.

Buddhism is not a religion but a philosophy of existence. Buddha is not considered a god-figure or godlike and he is not worshiped the way Christians are indoctrinated to worship "Jesus" (remember, that is not his real name).

Buddhist concepts and teachings are structured as a path of disciplined living which inevitably results in increasing individual realization and awareness. Buddhists believe that everyone has the realistic potential to experience the same self-realization and awakening that transformed a normal young man named Siddhartha Gautama into Buddha.

It is no coincidence that the fundamental concepts of Buddhism appear to be meant to bring the individual into closer union with the Universal Mind. For example, it is fundamental Buddhist teaching that there are three realms of existence:

1. Realm of Desire…our emotional reactions to sensory stimuli.
2. Realm of Form…the stimuli our five senses preoccupy our minds with.
3. Realm of Formlessness…that which exists beyond our sensory perception.

Buddhists believe most people exist in the Desire Realm of consciousness where experience is a constant source of craving and attachment to material things and their associated thoughts and sensory dependence. This craving and attachment to material "things" and sensory experience leads to unending suffering because the individual mind is untrained and becomes codependent upon these experiences. As a result, they cannot escape the Desire Realm. They are as tied to it as a thread is tied to a coat.

It is apparent that the Dalai Lama himself believes human consciousness is a consequence of the photonic-wave that has become the Universal Mind and that human Reincarnation is a reflection of reincarnated universes.

Why do we suggest this? The Dalai Lama has spent much of his adult life searching for proof of the connection between the Universal Mind and Reincarnation.

In a 2013 lecture in Tokyo, the Dalai Lama described his on-going investigation of how Buddhism and modern science are connected. He acknowledged engaging in a planned dialogue with scientists, concerning cosmology, neurobiology, physics (especially quantum mechanics) and psychology for over thirty years—most of his adult life. These conversations have been as beneficial to the Dalai Lama as they have been fruitful for the scientists he has met with over the years.

While the Dalai Lama learned a great deal about particle theory and matter, the scientists he has spoken with have begun to pay more attention to the inner world of the mind and emotions. As the Dalai Lama commented:

> "In order to understand the way our brains behave, we have to study our emotions and their effects. In the past, it seemed as though science and spirituality were opposed to each other. However, it's not a useful division to maintain…we need to know about both."

Source: Dalai Lama of Tibet, "Dialogue with Scientists in Tokyo on Universe, Life and Education," November, 2013.

Now why would the Dalai Lama spend over thirty years studying quantum physics, neuroscience (the science of brain development), cosmology and psychology? It appears that the Dalai Lama understands that all human individuality is reborn because human consciousness is part of the eternal Universal Mind. He is also aware that the Big Bang set in motion natural laws and relationships between matter and energy (photons or light) which make Reincarnation necessary to conserve our individuality.

As we proceed further, we will come upon more and more similarities between Buddhist doctrine and our theory that all consciousness derives from the photonic (light) wave emitted by the Big Bang which became coherent shortly thereafter.

As an example, consider this comment by the Dalai Lama on how Buddhism is centered upon the concept that the ultimate creative force in the universe is consciousness:

> "There are three different levels of consciousness. What we call innermost subtle consciousness is always there. The continuity of that consciousness is almost like something permanent, like the space-particles. In the field of consciousness, it is the **Clear Light…The Clear Light,** with its special energy, makes the connection with consciousness." (emphasis added)

Source: S. Rinpoche, The Tibetan Book of the Living and Dying, (interview of Dalai Lama), Harper-Collins, San Francisco, 2002.

The quote from the Dalai Lama, above, offers three significant similarities. First, I would translate the Dalai Lama's depiction of the Buddhist's "innermost subtle consciousness," as the Universal Mind. Second, his use of the term "space-particles" are, to me, synonymous with the photon. Third, the Dalai Lama's use of the term "Clear Light," capitalizing the first letter of each word to signify its importance. Then he adds that this "Clear Light" has a "special energy."

What light is the Dalai Lama describing? Obviously, it is not just any old light. This light is special. For one, it is "clear."

What does "clear" signify? In English, the term "clear" describes a substance that is transparent or translucent. You can "see" through it. But "clear" also means, "I understand. What you say is clear to me." By this meaning, the term "clear" refers to understanding or knowledge.

The Dalai Lama is describing a very special, "Clear Light" which possesses "special energy." What could that be? I believe we know—the photon-based Universal Mind. The Dalai Lama is connecting consciousness, light and information as one entity.

Not coincidentally, it just so happens that the photon-wave that became the Universal Mind was translucent; a very bright light that evolved into self-awareness as it absorbed all the knowledge released by the Big Bang and used to determine all the relationships between energy and matter during the early universe. With inspired intuition, the non-physicist Dalia Lama is describing the scientific composition of the Universal Mind itself.

But these are not the only similarities between basic Buddhist beliefs and our theory that the universe is pure consciousness, we are joined to it by photonic energy-waves and that Reincarnation is necessary to preserve our individuality.

Consider the Buddhist focus on negative emotions and their long-term negative effects reflected in the core of Buddhist teaching, **The Four Noble Truths**:

1. All existence is suffering.
2. The cause of suffering is craving.
3. The end of suffering is the end of craving.
4. There is a path that leads from suffering.

Each of these aspects of "suffering" involve memory as a function of consciousness. You cannot crave a thing you cannot recall. Memory as an expression of information and consciousness is, therefore, inextricably intertwined with craving and the Buddhist concept of suffering.

In my book ***REINCARNATION: The Scientific Theory***, I describe how individuality is a function of our past experiences preserved as memories in the brain. Evolution and natural selection determined that the strongest memories are those upon which we place the strongest emotions. Thus, the memories most likely to survive from one life to the next are those memories which hold the strongest emotions.

Emotion-loading of a memory imprints the memory upon the synapse like a photograph. The stronger the emotional-loading, the stronger the memory. The stronger the memory, the more likely that memory becomes part of your individuality and is preserved from life to life.

Unfortunately, evolution and natural selection have seen to it that negative emotions and negative memories are given much greater weight than positive memories. This difference was necessary for survival. Most people forget birthdays and anniversaries; but no one forgets stepping on a rattlesnake or the time they caught Covid.

In fact, I specifically point out in ***REINCARNATION: The Scientific Theory***, that one of the reasons people do not remember their past-lives is because the past-life memories that are more often preserved are negative memories built upon strongly negative emotions. This is very consistent with Buddhist thought:

"From a Buddhist perspective, it is important to appreciate the concept of emptiness. It is through the growing experience of emptiness that we experience a growing wisdom or awareness of our true nature **by eliminating all disturbing emotions** (emphasis added)."

Source: Tokyo dialogues (cited above).

The importance upon which Buddhist teachings give to the concept of "emptiness" is also significant regarding the Universal Mind. The practice of meditation, as a technique to achieve an evolving connection between the individual mind and the greater "emptiness," is an intuitive appreciation of the relationship between what humans perceive to be consciousness and the Universal Mind.

The Buddhist journey to "emptiness" is the process of eliminating all conscious attachments to negative emotions and the negative Karma they produce that draws us back into a rebirth into another life of craving and suffering. This concept is totally consistent with the concept that the memories that beget individuality are more likely negative than positive ones—due to evolution and natural selection.

The Dalai Lama's views in his books and lectures demonstrate an intuitive grasp of the existence of the Universal Mind and an inspired understanding of the relationship between the Universal Mind (coherent photonic-wave) and what he calls "subtleties of mind" or consciousness:

"Buddhist literature discusses the existence of different levels of consciousness, or subtleties of mind and how each level of consciousness corresponds to a different level of energy…most of our conscious mind consists either of states related with objects that we have experienced to the past—recollection of past experiences inform our present consciousness—or it consists of some kind of feeling or sensation. As a result, it is very difficult for us to glimpse the actual nature of consciousness, which is the sheer state of knowing or the **luminosity of mind** (emphasis added)."

Source: Dalai Lama, A Simple Path, p. 78, (cited above).

Here again, we see the Dalai Lama's intuitive link between "luminosity of mind" and the luminosity of photons and the photonic-wave. The connection between rebirth (Reincarnation) and the Universal Mind is ever-present in Buddhist cosmology. Rebirth is causally dependent upon the continuity or eternal existence of consciousness:

"Where does consciousness come from? Consciousness cannot arise out of nowhere. A moment of consciousness cannot be produced without the moment of consciousness that immediately proceeded it."

Source: S. Rinpoche, The Tibetan Book of Living and Dying, p.93, cited above.

The act of awareness requires a preexisting awareness. Unwind the string of causality far enough back in time to its logical origin and you arrive at—the Big Bang, and its creation—the Universal Mind. As the Dalai Lama expressed it:

"...Buddhism believes in universal causation, that everything is subject to change, and to causes and conditions. So, there is no place given to a divine creator, nor beings who are self-created; rather, everything arises as a consequence of causes and conditions. So, mind or consciousness comes into being as a result of its previous instants."

Source: S. Rinpoche, The Tibetan Book of Living and Dying (interview of Dalai Lama), p.94, cited above.

In other words, Buddhism is a union of individual spirituality and the search for scientific truth of where we humans "fit in" to the universal scheme of existence. If one can get past the Tibetan terminology, the basic structure of Buddhism is very simple to understand. Yet it is extremely difficult to master—because it challenges the individual to master the inner-self.

That may sound too overtly "religious" for some. However, I take the view that Buddhism is nothing more than a technique of self-mastery. Self-mastery is not a religion because there is nothing to worship. It is but a path of enlightenment.

Buddhism is only a solitary pursuit of inner-realization and self-knowledge. Buddhism, like it or not, is the closest humanity has come to a structured effort to achieve communion with the Universal Mind.

Don't get me wrong. One does not need to be a Buddhist to achieve communion with the Universal Mind. You do not need to BE anything to reach communion with the Universal Mind. Any time you are inspired by an image or idea beyond your experience, you have touched the Universal Mind—even if for just a brief moment in time.

As part of their sacred practice, Buddhist monks often create sand mandalas. The creation of a mandala can require the combined efforts of as many as twenty monks, working in shifts around the clock for ten days. To create one mandala, millions of tiny grains of sand are poured from traditional metal pipes called "chak-purs" over a period of ten days. Upon completion, following a brief ceremony, the mandala is destroyed, acknowledging the impermanence of all material things.

"A mandala is a symbol of the universe in its ideal form, and its creation signifies the transformation of a universe of suffering into one of joy. It can also be used as an aid to meditation, helping the meditator to envision how to achieve the perfect self."

Source: Loseling Monastery, "Mandala Sand Painting: Creating an Enlightened World, Copyright 1998.

Illustration No. 10. A Buddhist monk uses a Chak-pur to apply individual grains of colored sand onto a mandala.

Credit: normanculturalconnection.org copyright 2022.

CHAPTER 11

Mirrored Reflections

If the universe self-aware, how did this happen?

Before we attempt to answer this question, let's figure out what "self-awareness" is all about first. This is important. How we define the Universal Mind depends upon how we define "self-awareness" and its impact upon our relationship with the Universal Mind.

When I was a baby, my mother used to bathe me in the kitchen sink because every time she put me in the bathtub, I stuck my head under the water. It was fun. No one told me I could drown. The kitchen sink experiment did not last, either. I had too much fun playing with the little round, brown balls floating around. It made me giggle to see her discomfort as I played so naughtily.

I had fun. There was a definitive "I" about this reflection. I was having fun…not her.

In the middle of all those asthma attacks that were not fun, I was not having fun. I was physically suffering, not my parents or the doctors. I knew I was alone in experiencing it.

I was not having fun. There was a definitive "I" about this reflection, also.

Identity of self can occur at any age. All that is required is experience plus the capacity with which to observe and reflect upon this experience as separate from others. Something is happening to me that is not shared by anyone else. The rest is deductive reasoning which requires intelligence through evolution.

Any system is internally aware of its state of relative existence. But it takes intelligence to know there is a difference between self-experience and non-self-experience and to appreciate the distinction. As we exist over time, we observe events and emotions happening to us that are not shared by others. This realization probably occurs very early in childhood. Through this experience, one learns the self is separated from non-self—anything not identified as being of the self.

But this distinction of experiential awareness of self-versus-non-self requires a physical sense of separation. The universe is naturally all there is—there is no separation. Therefore, the universe lacks any experiential basis to form a sense of separateness from its surroundings.

Logic supports this conclusion. Anything that is a whole unto itself is, by definition, incapable of perceiving separation from any part of itself—there are no parts in its perception—just the whole.

But we also know that perception is relative. We humans, trapped in physical bodies that reinforce the conscious awareness of individuality know, only too well, what it feels like to be physically separated from each other and the rest of the universe. Prolonged isolation can be psychologically and even physically harmful to us.

Can we conclude that when the universe evolved into a Universal Mind, its frame of reference was that of a "wholeness" without any perception of physical separation from anything—anywhere?

No.

Remember the process of the Big Bang we discussed earlier. Two light gases, hydrogen and helium are experiencing super-heated and super-gravitational expansion in all directions, like a plasma ball. Surrounding this plasma ball is the thin layer of lithium that serves as a skin. Within it all, individual photons race outward.

Some photons are absorbed by the lithium, creating the universal-semi-conductor that stores the energy and information being held by the photons. Other photons, however, are being refracted, and reflected across the lithium skin back into the plasma maelstrom.

Refraction and reflection within the lithium sphere are both critical to how the universe became self-aware. Let's examine what each term means. Refraction is simply the bending of light waves as the waves pass from one medium to another.

Illustration No. 11. Refraction air to glass and back to air. copyright 2022

A light wave refracts or changes direction as it passes between different densities of matter. In the illustration above, a beam of light noticeably changes direction as it transits from air to glass and back to air again. Air is not as dense as glass, and the light wave will be altered as it passes through these different densities.

The early universe possessed a similar environment. Light waves traveling within the plasma were subjected to extreme variations in matter densities as they traveled from vacuum to plasma to gas to metal and back again—all the time being subjected to extremely unstable electric and electromagnetic fields and gravitational forces.

Most of us know that radio-waves are electromagnetic waves, just like light. This is true because all electromagnetic energy is light (photons). We can identify specific light energy by its wavelength and frequency.

Every light wave had a specific frequency and wavelength…

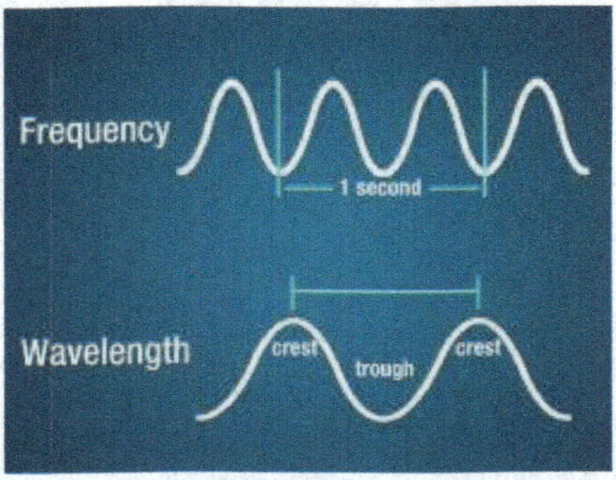

Illustration No. 12. Wave frequency and wavelength. copyright 2022

The frequency describes how much energy is stored within a particular wave of light. Gamma rays released by a nuclear explosion, for example, contain so much energy that they can be fatal to life. On the other hand, light from the Sun possesses much less energy and may cause only a burn.

The wavelength of light describes how long the wave is from trough to trough or from peak to peak. The longer the wavelength, the lower the light wave's frequency. In other words, the longer the wavelength, the less energy the wave possesses. Therefore, frequency and wavelength of light are inversely proportional to each other. Increase the energy of a wave (frequency) and you shorten the length of the wave. Decrease the wave energy and the wave flattens out.

Different types of electromagnetic energy (light) have different wavelengths and frequencies. Light waves have extremely short wavelengths (nanometeres). Radio waves have much longer wavelengths, varying from millimeters to kilometers.

Source: no author, "Wavelength," National Center for Atmospheric Research at scied.ucar.edu.

Refraction changes the energy of the light wave which alters both the frequency and the wavelength. This change is very much like what happens when you turn the channel on your television. Altering the frequency setting in your receiver allows you to tune in to another station.

The same thing occurs when you adjust the frequency dial on your radio. When you turn the dial up, it increases the frequency of the signal; when you turn the dial down, you decrease the frequency of the signal.

Each change in frequency allows you to turn in to a different station because each radio station has been given specific frequency settings for their transmitters to operate on by Federal Communications Commission (FCC) rules and regulations.

At the same time light waves were being refracted within the early universe, other light waves were being absorbed into and reflected off the universe's lithium skin. When a photon strikes an object, it will either be absorbed into the material it strikes or reflected away from it. If the surface of the lithium skin was smooth and shiny, the photons not absorbed would have been reflected backward in the path in which they arrived. However, if the lithium skin lost its smoothness, the light waves would reflect at angles.

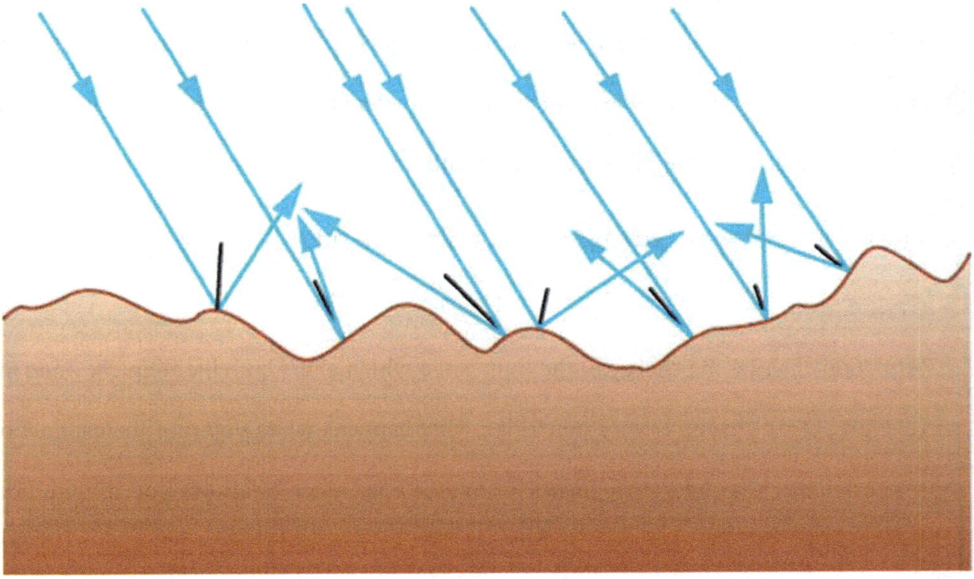

Illustration No. 13. Light reflection off both smooth and grainy surfaces. copyright 2022

The above illustration shows how light waves are reflected in different directions depending upon how smooth the surface they strike and the angle at which they strike it.

Source: Creative Commons. No known author.

Illustration No. 14. Laser light reflections. copyright 2022

The illustration above gives an idea of what it must have been like to be in the expanding ocean of photons inside the early universe's lithium shell. Each photon can be seen to interact with each of the others as they change direction within the lithium sphere. Angles of reflection would alter the trajectory of each photon, allowing the light waves to crisscross each other constantly.

Now imagine these photons all so tightly compacted together that you can hold 100 trillion of them in your fist and you have an idea of how closely bound together the photons were and how they could all share their information with each other so rapidly.

These crisscrossing photons collide. Each collision forces the photons to interact with each other within the highly charged electric and electromagnetic fields existing inside the lithium shell. The result is an ocean of intermingling photons.

This interaction creates quantum entangled photons. These now-entangled photons begin to share the absorbed information they possessed originally as separate photons.

At the same time as the photons are being entangled by crisscrossing reflection, they are changing frequency and wavelength up and down the electromagnetic spectrum through their refraction as they enter and exit different areas of relative density within the lithium shell.

Every time a photon is refracted, it is forced to change its state of being as well as its individual frame of reference. This must be so as the identity of each photon is contained within its specific frequency and wavelength. Frequency and wavelength are the only means by which one photon can tell itself from another. Changes in wavelength and frequency also change the information each photon possesses.

These changes alter the universe on a fundamental level. The universe is no longer a single entity encompassing non-interacting, non-experiential matter. The universe is evolving into a system in which intrinsic change is occurring all the time. Each movement of each photon evidences these constant changes.

Lastly, the lithium skin itself comes into play as it serves as both universal-super-conductor, computer memory and battery. Photons that have been entangled and have shared information with other photons eventually collide with the lithium shell. For every photon reflected off its surface, a certain proportion are absorbed—along with the information they possess.

The information within the absorbed photons, along with their quantum entangled unity is shared with the emerging consciousness being created within the lithium shell. We know from the Second Law of Thermodynamics that randomness increases with time and that, therefore, information cannot be destroyed.

The early universe was full of information being transferred from photon to photon as the mathematical relationships between matter and energy were determined during the universe's initial 380,000 years. As stated earlier, this occurred during the "decoupling era" when conditions were too hot and gravitationally oppressive for photons to escape the central universal core. These conditions imbedded the information within the "photonic wave" during the "decoupling era." At the end of the "decoupling era" this information was released with the photons that carried it.

Where did the photons and their information go? To answer this question, we must examine that lithium shell in more detail.

Every organism we know of has some form of structure. If the universe is self-aware, it qualifies as an organism (entity?) and therefore its consciousness has structure. We have already analyzed some of the components of the universe's structure (galaxies, connecting filaments of gas clouds, dark matter and photons, etc.).

Consider the following logical analysis: one of two conditions occurred during this time. Logic says either the universe formed a membrane, or it did not. The universe could do neither, nor could it do both. So, there are only two options:

1. The universe formed a lithium membrane.
2. The universe did not form a lithium membrane.

What is a membrane? Think of the skin of a bubble. It could be a soap bubble or a bubble from the ocean surf. A bubble is composed of gases emanating from whatever created it, encased in a thin layer of material.

If the universe possessed a membrane what would this membrane be made of? Any membrane consists of the materials it covers. A soap bubble is made of the oil in the soap. That is why its skin looks like a rainbow as light refracts through it.

In the case of the universe, a bubble membrane would consist of the original elements that were created directly from the Big Bang itself. As we discussed in an earlier chapter, the Big Bang produced only three elements. All matter released directly from the Big Bang consisted of 95 percent hydrogen, 5 percent helium and a trace amount of lithium.

Hydrogen and helium are the lightest elements in the periodic table of elements and exist as gases even at room temperature. Lithium, though a metal, is the third lightest element in the universe—so light it can be cut with a knife. The significance of lithium is that lithium creates and stores electrons—electrical energy. That is why lithium is used in making batteries for computers and vehicles. Lithium creates, carries, and stores electric current.

Source: NASA.gov.

There is your universe membrane—a bubble of photons stored in a lithium-ion shell.

As the "decoupling era" ended and molecules began to form, super-heated hydrogen and helium gases expanded and molded onto lithium atoms, forming an ever-expanding bubble of electrically-charged lithium. This membrane would absorb the information-rich photons from the photonic-wave that had just been released from the plasma-field and mix them with the ion-rich lithium. The lithium within the bubble then stored the released photons and, *voila*!

There you have it!

The universe became the first lithium-powered computer processor, with its own internal battery and infinite energy source. But it was not just any computer processor. The universe bubble now contained photons, which interact with each other in ways that even mystified Einstein with their "spooky" characteristics such as quantum entanglement.

There must exist a capacity within the mind to unconsciously interact with the "nonlocalized" quantum state created by the Big Bang's photonic wave. Remember, this photonic wave was emitted by the Big Bang and has immersed the entire universe within its informational matrix ever since—including all life here on Earth.

That is why we seem to feel a strange "oneness" with life whenever we receive inspiration that suddenly, without warning, replaces our everyday, classical mechanical thinking process with this nonlocalized, higher domain of awareness and the insight this "super-awareness" provides us.

Photons have properties we still do not understand; some of which turn our view of existence and linear reality on its head. For example, consider the quirky nature of quantum entanglement.

As we discussed in in Chapter 5, the first serious consideration of quantum entanglement was published in the now famous 1935 article in which Einstein remarked that quantum mechanical interactions created what he called, "spooky action at a distance."

Source: A. Einstein, B. Podolsky, and N. Rosen, "Can quantum-mechanical description of physical reality be considered complete?", Physics Review, 1935.

To illustrate what Einstein meant, consider the following scenario: two photons are made to come within extremely close proximity with one another by use of strong magnetic fields and then are made to fly off separately at the speed of light until they are billions of miles apart. Despite their extreme distance from one another, a quantum mechanical measurement of either particle will instantly determine the properties of its entangled partner. While this violates the speed of light, it does not violate the speed of thought.

Of course, this instantaneous communication between photons billions of miles apart violates the Hubble Constant which states that light can only travel 300,000 km/sec. So, photons possess capabilities we still do not understand. Apparently, photons can communicate with each other faster than they can travel.

For example, scientists still do not understand how the universe is expanding as fast as it is. Some suggest that dark matter has something to do with it. Now, I understand that photons have no mass and should not have any effect upon universal expansion…

But…

We know photons do interact with matter, indirectly. We call one of these interactions, photosynthesis. Photons combine with plant life to create oxygen. While we understand photosynthesis, photons have many other capabilities we do not understand, such as creating energy as well as absorbing, transporting, and sharing information.

These seemingly magical capabilities photons possess raise questions. For example, imagine a bubble-membrane of photons racing away from the center of the universe, with a core of physical matter, expanding from out of the center, but at a much slower speed. Such a bubble-membrane would contain all frequencies and wavelengths of electromagnetic spectrum energy. As this membrane would also include visible spectrum light, it would glow white—as it would contain all the colors at once.

But we have now come face to face with a universal-sized dilemma. This dilemma can be put as a question. The question is:

How does the universe expand when it travels slower than the photons it creates?

Is this true? If it is, don't we have to find out what happened to all the photons from the original Big Bang? Let's find out.

We have two objects expanding outward from the same origin. The first object is the entire, young universe. The second object (actually a set of identical objects) is the light (photons) emitted from the Big Bang no longer held back by the "decoupling era."

The dilemma is simply this: if the photons within the early universe traveled faster than the universe itself expanded, what happened to the photons? Where do the photons go?

First, is this true? To find out, we need to know two facts: (1) the speed of the expanding universe; and (2) the speed of light. Then we compare the two. If the speed of the expanding universe is faster than that of light, we have no problem. But, if light travels faster than the universe expanded, well, that is a problem, isn't it?

How fast does the universe expand? The rate of universal expansion (otherwise known as the Hubble Constant) is 73.3 km/sec/mpc (Source: Science News, March 2021). "Mpc" stands for megaparsec, and its scale may be a bit more than you can wrap your head around. So, let's take a look at the relativistic speeds of cosmic distances easier to understand than a megaparsec.

The speed of our Solar System is approximately 720,000 to 828,000 km/hour (Source: Physics Factbook at hypertextbook.com). The speed of our Milky Way galaxy is approximately 600 km/sec (Source: astrosociety.org).

Compare these speeds with the speed of light—300,000 km per second in the vacuum of space. It takes our Solar System an entire hour to travel the distance photons travel in just two seconds. Our Milky Way galaxy takes 8.333 minutes to travel the distance a photon travels in just ONE SECOND!

Why are these speeds important? The universe consists of two structures which carry gravitational weight relative to other matter—solar systems and galaxies. Both have tremendous inertial momentum. The only universal structures which would have any significant impact upon their relative speed are other solar systems and galaxies (structures of similar mass).

Therefore, it is reasonable to assume that the expansion speed of the universe relative to solar systems and galaxies is relatively equal to our own Milky Way and Solar System.

There is no evidence to suggest that our Solar System or Milky Way are significantly different from any other in terms of relativistic velocities.

These facts lead to several conclusions. First, the velocity our Milky Way and Solar System travel relative to other cosmological structures is an accurate indication of the universal expansion rate. Second, the universe is clearly traveling (and therefore expanding) much slower than photons travel.

Why this is so, is simple. When the universe began expanding at the end of the "decoupling era," both matter and photonic energy were released with great force. Newton's Laws state that once a mass is in motion, it will remain in motion until an opposing force slows it down. The matter released by the universe at the end of the "decoupling era" would have remained at the same speed—unless some opposing force existed.

Therefore, the speed of the universal release of matter at the end of the "decoupling era" can be reasonably estimated by the current speed—as there is no evidence any significant gravitational forces opposing the released matter interacted with this matter. There was no opposing force relative to the matter released. Therefore, we can reasonably conclude that the speed of the Milky Way and our Solar System gives us a reasonable estimate of the force exerted by the "decoupling era" expansion.

Physical-particle matter has weight and mass. Photons, on the other hand, are massless and weightless. The expansion rate of the universe is much slower than the speed of light because the matter inside the universe (both ordinary and dark matter) holds the expansion back by relativistic gravitational forces which do not interact at all with photons.

So, when the early universe transitioned out of its heat and gravitational plasma-field at the end of the "decoupling era,"—where did all the original photons go?

We know these photons traveled orders of magnitude faster than the rest of the universe. The photons flashed ahead, leaving the physical matter of the universe further and further behind.

What does this mean?

Wherever the present universe we are familiar with goes—that of galaxies, gas filaments and dark matter—photons have already been there—millions, perhaps even billions of years earlier. Whatever region of space our Solar System travels, photons from the Big Bang have already been there—millions, if not billions of years before. We are forever traveling through a space saturated by photonic energy for billions of years.

Therefore, it is likely that our universe consists of an expanding wave of coherent photonic energy embedded in a lithium bubble billions of light years ahead of the matter-portion of the universe (galaxies and dark matter, etc.) and that in between the two exists an area of lower relative density—relatively barren of matter.

These two varying areas of pressure density may explain universe's mysterious expansion rate. The lower density of the empty space is pushing the lithium skin ahead of it outward as is "sucks" the matter-rich, higher pressure density part of the universe following behind.

At some point after the universal lithium-photon bubble formed, the merger of quantum-entangled photons and the lithium-powered electromagnetic field became self-aware. The emerging universe had just given birth to a Universal Mind.

CHAPTER 12

THE UNIVERSAL MIND

I would like to take you by the hand and lead you on a journey into the Universal Mind. But I cannot. You are on your own. Each of us must assume responsibility for our own connection with the universe. Your truth, your unique individuality would be squandered if you chose to leave it behind by following in someone else's footsteps. What I can do, however, is share my own truth with you and hope it sheds some light upon your own path.

To do this, we must go back in time to when I was only six years old in Phoenix, Arizona. It is 1956. I had just had my near-death-experience in the hospital emergency room. We lived on E. Pinchot Avenue, next to E. Thomas Road.

Back then, E. Pinchot Avenue was THE last street between Phoenix and Camelback Mountain. Today, the land between Phoenix and Camelback Mountain is the Phoenix suburb of Scottsdale, Arizona. But back in 1956, the only thing between my back yard and Camelback Mountain was raw, untamed desert.

Following my Near-Death-Experience when I was just 6 years old, I spent the next two weeks recovering at home. Voices in my mind kept telling me to go to Camelback Mountain.

I spent hours in our back yard, just staring at it. I was obsessed by its majesty and hidden power. One day, when I went out to the backyard to stare at the mountain again, my grandmother was waiting for me. She was pretending to be hanging laundry out to dry in the Sun, but I sensed she was really waiting for me. She didn't take long to get to the point. She asked me straight away why I spent so much time staring at the mountain.

I wasn't sure how my grandmother would respond, but I told her the truth. I told her that ever since I had come home from the hospital, I had been drawn to the mountain and that voices in my head were telling me to "go to the mountain."

I expected her to scold me and warn me not to go. But there was a lot to my grandmother I did not know. Instead of scolding me she told me something about herself, she had never told me before—or to anyone else, I believe. My grandpa was inside. Mom was working. It was just the two of us, in the backyard, together alone.

She told me she was going to tell me a secret and never to tell anyone else.

"Sure, granny. I won't tell anyone."

She confided she was half-Native American and that her mother was a member of a tribe of Native Americans that had been living near Rising Sun, Indiana. Her father was a hunter-trader from Canada who spoke French-Canadian. I have since forgotten the name of her people.

Instead of warning me not to go to Camelback Mountain, she told me her people did not call the mountain by the name "Camelback." In fact, she was rather indignant that the mountain had been given that name. This was not like her, because my granny was one of the nicest, warmest, most caring people I have ever met.

But she was intensely offended by how mountains that were spiritually important to her people had been bastardized by Americans. According to my grandmother, the mountain had a different name among her people. I do not remember the native name she used but she told me it means “Woman Sleeping While Pregnant.” She also told me that among her people, the was a place of great spiritual power. That is all I needed to hear. I had to go to the mountain. She knew it, too. But she never told me not to go.

Illustration No. 15. Camelback Mt. This is about what it looked like to me from my backyard in 1956.

The photograph above gives a very good view of Camelback or “Woman Laying Down While Pregnant” Mountain. The tallest peak in the middle would be the belly of the sleeping woman who is laying on her back.

My mom was not impressed when I told her. "Don't go wondering around the desert," she scolded. "If the snakes and the scorpions don't kill you, the mountain lions and coyotes will. You stay in the back yard where you belong. Sick kids don't go off in the desert."

Her warnings went in one ear and out the other. From my backyard, Camelback Mountain didn't look that big and it sure didn't look that far away. If I ducked out of school during the afternoon recess figured I could hike there and back before dark.

I was afraid of one thing, however. I feared chickening out. I was scared to death I would find a reason not to go. So, to keep myself on the path of discovery, I asked my best friend, Mark, into going with me.

I didn't expect him to agree to come along. It would get him in trouble for no good reason. We would be skipping out of school. We would be doing something that would put us both at great risk.

"Yea, I'll go."

I was amazed.

"When are we going?" He asked.

"Tomorrow." I told him. I was afraid if I put it off, I would find a reason not to go at all. I would have hated myself the rest of my life if I had chickened out.

"As soon as we get out for 2:15 recess, we go."

Mark and I skipped out of elementary school at the last recess. We just took off and ran to my house. This was a risk. Grandma was no problem. But my grandpa would be a problem. He'd ask us why we weren't in school.

To my surprise, though, no one was home. I grabbed us a few sodas and canned bean soup for something to eat on the way. I even remembered to grab my mom's can opener and a steak knife—just in case we came up on any mountain lions.

Then we were off and running out the back yard and into the desert. "How many kids skip school to go to a mountain?" I asked. Mark shook his head. He couldn't think of any. I was very proud of myself. Mark, not so much.

"We gotta get home before dark," he said frowning. "My Dad'll kill me if I'm late for dinner."

"Don't worry," I assured him. "We'll be home way before dark. Imagine! What kid has ever skipped school to go to a mountain?" I congratulated him. "And we're only six!"

For a while, the two of us had a great time. We jumped over rattlesnakes, stomped on scorpions, and hid from a pack of coyotes hunting for food. Want to know what it was like to come upon a rattlesnake at six years old?

Mark and I were both so excited about the journey that we were running. As I'm running I just by luck see the rattlesnake curled up on a rock right in front of me; its rattler pointing straight up into the air. I just took one step and leaped as high as I could as I ran over it.

The experience made me high. I got giddy and started whooping it up loud as I could. I was ecstatically happy to be on the way to my mountain. But I was also thankful that I had been able to see the snake in time to jump over it. I would not want to have to be out in the desert in the dark, at night. That would not be very good. That would not be very good at all.

I came across a lot of scorpions and lizards, but fortunately, we never came upon any mountain lions or coyotes. After an hour or so, we ate our ration of bean soup and soda, surrounded by cactus and desert sand.

Then we went back to running to the mountain. I was having time of my life. In fact, it seemed that time stood still.

"Hey, stop," Mark called from the rear.

I stopped walking and waited for him to catch up.

Mark pointed back where we came.

"Look!"

There was nothing to see but desert…and tiny white lights flickering on and off in the distance. There was a chill in the air.

"What do you think those lights are?" He sounded agitated.

"I don't know."

"That's our neighborhood, moron," Mark said, giving me a backhand smack on my arm. "We walked so far, we can barely see home, and we're still not any closer to your dumb mountain. You said we'd be there and home by now!"

I had been so focused on my running that I hadn't looked at the mountain in what seemed like hours. I looked around.

Sure enough, Camelback Mountain still looked as far away as it had when we started.

This Photo by Unknown Author is licensed under CC BY-SA

Illustration No. 16. Arizona desert with Camelback Mt. in background. copyright 2022

My heart sank. How could I have made such a stupid mistake?

"Don't you know in the desert air everything large looks closer and smaller than they really are?" He yelled as he punched my arm, angrily. "You idiot! You brought us out here for nothing! NOTHING! My Dad's gonna kill me and it's all your fault!"

I tried to think of words to make it right with him. But none came. The mountain appeared so close when I was standing in my back yard. How could I have misjudged this so badly? I had thought I had planned this trip out. How could I have made such a stupid mistake?

"I'm sorry," my head lowered. "I shouldn't have talked you into coming. I was just scared I'd chicken out if you didn't come along. I had to go…"

"Look at the Sun! Shit, just look around!" Mark waved his arms in a circle.

I looked up. The sun was setting. Violet, pink and orange-colored swirls streaked across the sky. Below, dark shadows crept toward us. Behind us, off in the distance, were the lights of home, twinkling like little stars.

"It's almost dark."

"That's why the lights are on, dimwit. It's dinner time." Mark said, shaking his head. "It'll be dark soon, and we won't be able to see. How do we get to the mountain if we can't see?"

"I don't know."

"Well, I'm goin' back." Mark was resolute. "Come back with me. I don't want to go back by myself."

"I can't."

"What do'ya mean, you can't?!" Mark punched my arm again. "There's no point to this anymore! If you keep going, you'll end up dead. You know that, right? Let's go back now." He stood there, staring me down, the red sun shining in Mark's eyes.

"I just can't."

"Why the hell not?" He demanded.

I told him about almost dying from asthma the month before. I told him everything; from seeing the disembodied head of a woman floating outside the family car on the way to the hospital to seeing the portal open and the white orbs floating around and even the moment the doctor plunged the syringe into my heart as I watched from up above the trauma room lights.

“I have no choice,” I answered, shivering. The night air was getting colder. Goosebumps were crawling all along both arms. “If I don’t keep going, I might as well have died at the hospital.”

“I didn’t know,” Mark rubbed my arm where he punched it. “Sorry. You ok?”

“I’m ok,” I replied, trying to hide the wince as he rubbed. “Just tell your dad it’s all my fault.”

It was crazy. It was insane. Two six-year-olds in the middle of the Arizona desert and we were completely nuts. And it was completely glorious.

It was the most ridiculously incredible moment of our lives. And we realized it together. We had stepped through a magical rift through time and space into a region of existence very few ever touch. Whatever happened to us, we both knew we had done something very few dare to do.

It was special. More than special, it was unique—one of those rare “lifetime moments” that you never forget—and it was happening to us together. Everything suddenly made a crazy kind of cosmic sense.

We laughed until our sides hurt.

“I gotta go.”

Time stopped. Mark looked at me, all his anger gone now. In its place, a resigned sadness took root and grew upon his face. We both knew what came next.

“Come back with me,” his eyes watery. “Please.”

What could I say? My head shook, slowly…silently…from side to side. With a half-hearted wave, he turned and started walking back where we had come. I watched as he slowly disappeared into the shadows.

As Mark disappeared, I felt the urge to run after him. A part of me wanted to go back home with him. The tiny lights of my neighborhood called out to me to hurry back before dark so I could be safe and warm. Best to go home with him now, before it got dark. Before I became prey.

I better run after him. Now! The urge was almost overwhelming.

Run!

But I stood unmoving as though rooted to the desert floor.

I was terrified to go on.

But I did not move.

My 6-year-old mind screamed to run after Mark and go home with him.

But I did not move.

With an almost inhuman strength of will, I choked down the rising panic in my chest and fought the blind urge to run back home. Then, turning my back on the lights of home, I started walking toward Camelback Mountain.

My choice had already been made. It had been made for me weeks before, in a hospital emergency room, while my consciousness floated high above the lights in a trauma room; the day I almost died...

But there is no "almost" to death. Those of you who do not recall your past-life deaths believe consciously that "Death" is a door that opens only one way. You think that when it's your time, you go through the "Death-door" and you don't come back.

Well, let me tell you that you are wrong. You can come back from the "dead." I know, because it happened to me.

There is no "almost" to death. Surviving a NDE is like diving into the surf on a beach. You step out of the surf on the beach—but you can never dry off. The "water" never dries. It sticks to you the rest of your life. You can never forget what it was like floating in-between the physical life here and the "other side." You are forever changed.

****

There was very little daylight left to see with. I had not thought to bring a flashlight with me. There had been no reason to bring one. We were supposed to be back home by now.

A sudden chill crawled up my bare arm. I knew the desert between my back yard and Camelback Mountain was a dangerous place; filled with poisonous snakes, coyotes, mountain lions, cougars, and wolves—just to name a few.

Smart or not, I had no fear of any of them during the day. I believed them all to be night predators. The heat of the day would keep them in cool burrows and caves. But as the darkness fell upon me, I knew my risk of death from predator attack increased with every minute I remained out in that desert.

I began walking as quietly and as quickly as I could toward my mountain. I traveled as though my life depended upon it.

“Do not run,” I kept repeating to myself. “Do not run.”

Only two animals run in the desert night—predators and prey. I had no chance of being the former and no desire to become the latter.

Source: (no author, no date) Creative Commons.

Illustration No. 17. Camelback Mt. at sunset.

Night fell quickly—too quickly for me. Nightfall in the Arizona desert is surreal. The heat trapped in the desert floor begins to float upward. The cold air above me mixed with the rising heat from the desert floor below. After that, my mind kind of went blank. I just kept walking.

Once the sun set, the desert floor became a predator of its own. It was huge and dark with the shadow from Camelback Mt. I could no longer see the desert floor, so I began shuffling my feet. If I kept walking or running, I risked tripping over a rock or cactus. Worse, I might step on a rattlesnake. That wouldn’t end well.

I had no time to think about any of this, however. Like a monstrously dark hand, the shadow of Camelback Mountain enveloped me in its grasp. I could no longer see anything on the ground—not even me. I was swallowed up amongst the blackness; all alone in the desert night.

But fortune smiled on me.

It was a cloudless night. Even luckier, I had chosen to make my journey on a night with a full moon. Below me lay only shadow and the rising breath of heated desert floor gone deadly cold in the night air. But above me, the black, foreboding night sky began to slowly come alive with stars.

To manage the fear of what was below me that I could not see on the desert floor, I focused on the stars that began popping out of the dark sky above. To see each new star pop out of the darkness above brought me immense joy. Each new star became my newest and best friend. Each new, best friend brought me renewed hope and courage to keep going. Pretty soon, the sky was filled with my best friends.

The night air became very cold. My arms and legs quickly began shivering and filling up with goosebumps. Not expecting to be in the desert at night, the only clothes I had were the ones I was wearing—a T-shirt and khaki-shorts.

The cold became an increasing worry for me. Very quickly I began having to spend as much time rubbing my arms and legs to keep warm as I was taking to walk.

Getting to the mountain and going back home that night were no longer on my mind. My dream of reaching Camelback Mountain had been replaced with the reality that I probably would not survive the night.

I kept going forward.

As I walked, I developed my own survivability plan. My best possible chance of surviving the night was to get as close to the mountain as possible. The open desert and the cold night air were my new enemies.

Any hope for survival existed in getting close enough to the mountain so whatever heat that rose up from the mountain would keep me alive. Hopefully, there would also be some sagebrush or something that I could use to hide myself from predators and provide some warmth.

It sounded like a good plan. But, first, I had to get to the mountain.

Trekking across the moonscape that was the Arizona desert, the stars became my friends. I lost myself in the darkness. The Arizona desert disappeared. Instead of walking to a mountain on Earth, I was walking on the moon or another planet. I was alone, drifting in outer space.

There was no longer any point to looking at the ground, so I focused solely on watching the stars above me. My eyes focused solely upon the starry night sky. Time stopped. The desert was no longer a desert, but a universe and I was its only explorer.

Under the stars, alone, I owned the universe. My mind detached itself from my body. I became a disembodied spirit, floating in the cosmic night. I felt a communion with whatever governs existence that transcended space/time.

I fell face first into the cold, hard desert floor. My face and knees slid along the gravel-like sand. Flesh ripped away from my cheek and both knees. The sand in the wounds stung. My communion with the universe had ended in a skidding heap, scraped and bleeding.

I had tripped over something I couldn't see in the pitch-black darkness. Rolling over, I tried to sit up; then collapsed into a ball of uncontrollable shivering flesh. My entire body had shut down. My muscles spasmed uncontrollably. My teeth wouldn't stop chattering. A presence sat down next to me.

Is this how it ends?

MOVE!

I began rubbing my thighs and calves as hard as I could to try to get some feeling back in them. The feeling in my legs came back enough that they stopped convulsing.

I stood up. The night air had turned very cold. My skin immediately began to goosebump. I couldn't stop shivering.

I had run out of time.

Panicked, I began running as fast as I could. It was noisy, and I was making myself a target for the predators. But I no longer cared about them. Running was my only hope to survive. If I didn't get some body heat, I was dead anyway.

I didn't run far. My leg muscles spasmed again. I fell face first, skidding upon sand and rock. My knees screamed. My cheek and nose stung.

I tried to move my hands to rub my legs, but my body refused to cooperate. My arms and legs shivered uncontrollably. I was incapacitated.

It's over.

Curled up like a ball of quivering flesh on the desert floor, I prepared to die. Soon, what felt like death curled up beside me like a warm blanket of soothing serenity. I felt no fear, merely an acceptance of the inevitable. Having "died" once already, I knew the drill.

"Pick up a handful of sand."

A voice, but not a voice. A voice inside the mind…but not my own.

It's like someone inserted a microchip in your brain that connected you to a frequency you did not use yourself—like a TV channel or radio frequency. The second you "hear" it inside your head, you know it's not your own thoughts. To respond, you merely "think" the words and the communication is shared instantly.

Spooky?

Like Einstein's "spooky at a distance," there is a form of entanglement between you and whatever it is you are communicating with.

Who or what was it?

"Hey, I'm freezing here!" I yelled. "Give me a blanket or something, please!"

"Put the coldness aside. Pick up a handful of sand."

A bright flash of light blinded me. It was so bright, I had to blink until the light faded. But the flash of light had left me a gift.

The intense cold was gone! The uncontrollable shivering was gone. I was warm again.

"Pick up a handful of sand," the voice commanded.

I was no dummy. By this time, I knew I was experiencing telepathy. There was no audio coming into my ears from the outside. This was unmistakably telepathic.

But there was no sense of fear. My Near-Death-Experience had opened my mind to mental communication with other beings. This was no ordinary encounter with another human. Despite my youth, I was smart enough to appreciate the significance of what was happening. It was special.

No, this was beyond special. This was unique.

My fingers dug into the desert floor. I was surprised. The top layer of desert was not what I expected. It wasn't soft, like the sand you play in on a beach. In fact, the desert floor was very hard.

Using my fingers as a small claw, I raked them across the top layer of desert until I could get enough sand to fill my small hand. Finally, with sand in hand, I stood up.

"Open your hand. Let the sand fall through your fingers."

My small hand and tiny fingers opened just enough to let sand slip through them and fall onto the desert floor. The moonlight reflection made the individual grains of sand look like magic dust.

"Look up at the sky," the voice ordered. ***"Do you see the stars in the sky?"***

"Yes," I replied. Grains of sand still falling from my hand.

Source: Creative Commons (no author).

Illustration No. 18. Stars in your hand. copyright 2022

"There are more stars in the sky than grains of sand in this desert. Every star has its own planets. Imagine how many stars have planets with life. Yet, no matter how many stars there are, or how many planets have life, none are any more or less special than you."

A shiver ran up my spine. Pure joy spilled its contents over my head, filling me like a cup overflowing with infinite happiness. I laughed and cried at the same time.

"Will you remember this?"

"Oh, yes!...yes!" I cried. But I had begun shaking again. Whatever the white light had done to warm me up had worn off.

An image appeared in my mind of me covering up with brush for warmth. I knew this image was from the "alien" presence within me.

"You'll be found before dawn. Whether you are found alive or dead is up to you."

Then it was gone. I felt it leave. How did I know?

****

If I hold your hand, you can feel the pressure of my hand squeezing yours. Take that image and apply it to your consciousness. You know when something "alien" to you comes into contact with your consciousness. We all are familiar with our own thoughts. We instantly are aware of the difference when an "alien" presence begins telepathic communication. You can feel the slight change in the pressure within your mind. When the "alien" presence leaves, you can feel the void caused by its departure. It is a change in density.

Why is this so difficult to accept? Paramecium and other single-cell animals exist within membranes that are semi-permeable, meaning that some materials can pass through while other substances cannot. Evolution has changed the look and shape of membranes from that of the single-cell animals to the human skin. But both remain semi-permeable.

If medical science can use microscopic instruments to "pierce" the membrane of a woman's egg and artificially inseminate it, why can't a much more evolved species "pierce" the membrane of our consciousness for telepathic communication?

****

Grabbing what brush and scrub I could find, I made a makeshift mattress and laid down on it. I know the voice wanted me to make a blanket, but that proved impossible. The brush "cover" kept falling off.

I laid down and stared up at the stars the voice had talked to me about. Staring back down at me were billions of stars and millions and millions of life-possessing planets. Were beings living on those planets looking up at their sky at this very moment—looking up for me as I looked up for them?

I sat up, laughing and sobbing at the same time. Tears fell down my cheeks. I sat howling into the night. Predators no longer frightened me. I had become one.

In that moment, I knew why I was there. I was never meant to reach the mountain. Something in the universe wanted me to go out to the desert to have this encounter; to experience this moment.

Why?

Isn't that what our lives are all about—the "why?"

What was the point to it all if I was going to die? It made no sense to me. Then, in a flash of illuminated realization, came insight.

"Death" as we have been taught, does not exist. My previous Near-Death-Experience had shown me that "dying" did not end anything that really mattered. My body might die, but the knowledge and memory of this moment would last through and beyond infinite lives to come. It would stay with me forever.

Laying back down, I curled up into a ball and closed my eyes. My consciousness sank beneath the waves into the abyss. When I reached the primordial depths of time and space. I did not resist. The universe could do with me as it willed.

WHOOP…! WHOOP…! WHOOP…! WHOOP…! WHOOP…! WHOOP…!

A noise, louder than I'd ever known, lifted my mind up from the depths of unconsciousness. Opening my eyes, I was blinded by a very bright light. Something or someone grabbed me roughly and picked me up. It had to be some alien. I began hitting and kicking it as hard as I could.

"Stop it, dammit! If you don't stop, I'm gonna drop ya' and leave ya' here! You can freeze to death for all I care!"

The voice was human. When his head blocked out the blindingly bright light, I saw he was wearing a military flight suit. I stopped hitting and kicking him.

The bright light and the "whooping" sound I heard were not alien, but spotlight and rotor-wash from a Sikorsky H-34 military helicopter. My "alien" captor was an airman from the military base just outside Phoenix. On the flight back I got my first-ever cup of hot coffee—and a lecture.

"Lucky we found you, kid," the pilot said sternly. "Gave up on finding you. By the way, your mother's been crying for hours at the base. How do you get off making your mother cry like that? You little brat."

"We were on our way back to base when we saw your flashlight," his crewman added. "Been lookin' for ya for hours. You're real lucky. We figured ya'd gotten yourself chewed up by one of the critters out here."

I didn't know what light they saw. It certainly never came from me. There was no flashlight. I didn't think to take a flashlight along with us. Mark and I had expected to be back well before dark.

I was very wrong. What I thought was a small mountain close to us, was a much larger mountain that was farther away. No one told me it would take that long to get to the stupid mountain. But then, I never asked. It probably wouldn't have stopped me anyway. I would have just skipped school and gone during lunch.

As for that "flashlight" the helicopter crew told me they saw that told them where I was? I never have figured out what it was. One thing is certain. It wasn't me. I was unconscious.

Whatever that light was; whomever caused it to flash; saved my life. Unless you believe in little green aliens, there can only be one answer—the Universal Mind.

When we reached the military base, I learned Mark had managed to get home and tell his dad where'd he'd been. His dad called the cops. My folks had already called the cops when I never showed up from school, but they had no idea what happened to me. My mom thought I was dead. I never saw Mark again.

Thirteen years later, I am nineteen and working at a pizza parlor in Ohio. During my lunch break one day, I sat leaning up against a garbage dumpster with my slice of pepperoni pizza and soda lunch. The sun was very bright. Suddenly, I felt sleepy. I put my slice of pizza down on my lap and closed my eyes.

But instead of falling asleep, I found myself in deep space. I had no space suit, no rocket ship. Just me circling the planet Jupiter, its angry red eye-storm swirling below me.

The Sun was a pale, yellow ball below me toward my left. Where was Earth? I had to strain to see it. But there it was, a shiny blue-white dot in the blackness of space.

A presence approached me from behind. This was a different presence from the one I encountered in the Arizona desert. For one thing, it lacked subjectivity and any recognizable emotion. It was coldly objective. Together we stared at the fragile, little planet Earth, so far away.

Images appeared of human suffering; people in undeveloped countries with little or no food holding their young children with bellies unnaturally distended from starvation; crowds of people being slaughtered in the streets by black uniformed soldiers hiding their identities with balaclavas; expressionless soldiers arming on both sides of an snow-covered border as artillery and missile launchers ready for war; young people cowering in fear of an uncertain future; the elderly cowering in fear as the end of their lives inexorably approaches.

The cold, emotionlessness evaporated.

In its place was a feeling of overwhelming empathy and compassion. The depth of feeling was overwhelming. How could anyone be coldly objective one moment, but so compassionate and loving the next? The contrast seemed impossible.

Yet there it was. The presence that had come to me felt an immense compassion for the frightened, souls on Earth, so lost and alone in their ignorance of what exists in the universe so close at hand. It felt that we humans are frightened, lost children who have forgotten who we are. Because we have forgotten who we are, we rush around blindly; panicked, not knowing whether to laugh or scream, love or slay each other.

Then, like a feather in the wind, it was gone.

My vision of being alone in space vanished with it. I was back again, leaning up against a dumpster, behind a pizza restaurant. The sun was shining brightly. A slice of pizza in my lap.

"There are more things in heaven and earth,
Horatio,
Than are dreamt of in your philosophy."
Hamlet, Act 1, Scene 5 (Wm. Shakespeare, 1603.)

There are those who believe that we are not alone in the universe. However, too many of these well-meaning souls believe that all that's needed to attract the universe's attention is technological advancement. They believe that the sooner we have a Moon base, a Martian colony, corporations shuttling ore carriers to and from the asteroid belt, the sooner we will finally be contacted by whatever alien species exist in this part of the galaxy.

I do not share this view. Contact with the universe has already occurred. We are in contact with the universe every moment of every day of our lives—and we take it for granted.

I will not pretend to guess whether Earth-governments have or have not yet been contacted by alien species. Such questions are irrelevant to the real issue facing us as individuals.

Do not depend upon your government to intercede for you in your relationship with the cosmos—or save you from the world you helped create. The world is what it is because you helped let it become this way. Either through your actions or your inactions, your thought created the Karmic inertia threatening our Earth.

Are you an individual? Have you freed yourself from the fears and ignorance of the mindless-group? If you have, then you no longer require any agency to act on your behalf to meet the consciousness of the Universal Mind. You can do it all by yourself.

America owes its ideological inspiration to our Forefathers whose dream was an Enlightenment that would allow the Common Person to govern themselves without the need for royalty and dictators. The Enlightenment still exists as a beacon of hope that humanity can and will rise above the need for authoritarian rule with an educated populace.

That dream remains a distant vision, however. Until the nations and people of Earth resolve their conflicts over political power, religion, climate change, personal freedom, race, and economic equality, it is not ready to share in any open dialogue with another race. How can we hope to communicate rationally with an interstellar species when we cannot even communicate rationally with each other here on Earth?

Thoughts are things. They move us to speak and to action. Thoughts compel us to be or not to be who and what we will become.

Thoughts have power. No one can do anything until they think of it first. Turn that around and you realize that if you can think it, you can do it. The first step in turning any dream into reality is first to dream. Mental imagery is the magic wand. Consciousness is the wizard.

Stop making your god in your own limited image of yourself. Stop defining your god by your own prejudices, conceits, failures, and neuroses. Stop defining your god by your life experience. Go beyond it.

The Universal Mind exists as a purely conscious, nonphysical entity. Thought is the speech of consciousness. But who can listen when they are themselves talking all the time?

Can you? I don't think so.

Neither can I. We can only listen to others when we stop talking ourselves. The Universal Mind is no different. Can we expect to hear the universe speak to us when we allow our minds to be full of clutter and the background noise of our routine, daily lives?

This is why gurus and monks seek silence. They seek mental discipline to listen and hear the voice of the Universal Mind reach out toward them within their own minds. This is why anyone can experience inspiration while taking a simple shower or going for a walk alone.

When we shower, our minds become focused on a single thought—self-cleaning. Everything else we had been concentrating on, disappears. The mental clutter interfering with our communion with the universe dissolves like so many soap bubbles. We can finally listen.

When athletes compete, they exert themselves, push themselves to extreme levels of mental focus and physical performance. Every so often, their conscious selves leave their bodies. They perform their role; they can pitch, catch, run—all the physical activities necessary to compete, but their minds are detached.

This experience is called being in the "zone." Runners call it "runner's high." Both are the same experience. Once you experience it, you never forget it.

From personal experience, the "zone" is an effervescently fragile mental state. Once you are "in" the 'zone", your mind feels expanded in all directions, your awareness reaching outward toward infinity. But if you dare think about what you are experiencing; if you engage your self-awareness; the "zone" is lost. It dissolves like a soap bubble popping in your hand.

The trick is not to think, but to blank the active mind and surf the wave of higher consciousness wherever it leads you. It is an extremely intoxicating experience. It is also an initial, superficial step toward communion with the Universal Mind.

CHAPTER 13

What is a Universal Mind?

We all want to know what the Universal Mind is. So, let's take it a step at a time and see if we cannot gain some understanding of its basic nature. Let's use the Internet "Cloud" as an example. Imagine you are at your home personal computer and you need to "store" your data safely. You use the "cloud." What is the "cloud?"

"Simply put, the cloud is the Internet—more specifically, it's all the things you can access remotely over the Internet. When something (data) is in the cloud, it means it's stored on Internet servers instead of your computer's hard drive."

Source: "What is the cloud?", GCFGlobal, edu.gcfglobal.org, (undated) (no author).

Internet web servers are computers that store data as memory. How do they do that?

"The cloud storage uses a chain of servers that includes both master control server and other storage servers. The servers are all linked with one another, and it can be utilized depending upon your use and requirements and billed accordingly…The most fascinating feature of cloud storage solutions is you can upload your files or download them from anywhere across the World even without carrying your laptop.

You just need Internet access and any mobile device on which you need to download the application of your service provider. Then, simply login to your account and upload or download your files."

Source: "What is Cloud storage and how does it work?", Hitesh Jethva, Alibaba Cloud Community, alibabacloud.com, (2019).

The Internet "cloud" relies upon computers that use a traditional, classical mechanical method of computing—electricity organized as a series of digital 1s and 0s in computer formatted language.

On the other hand, the Universal Mind is made up of photons which operate according to non-classical mechanical rules called quantum mechanics. The universal "cloud" was created by the original flash of photons released at the time of the Big Bang. These photons washed across the new-born universe countless billions of times, saturating it with their light and information—all the information released by the Big Bang.

First, we need to remind ourselves of what we went over in Chapter 3. Photons and atoms of hydrogen, helium and lithium were the only particles created by the Big Bang. Hydrogen and helium combine to form helium hydride. Believed to be the first molecule formed after the Big Bang, the presence of helium hydride in the interstellar medium was confirmed in 2019 from the SOFIA telescope.

Sources: B. Andrews, "Scientists Find the Universe's First Molecule," Discover, 2019; and R. Gusten, H. Wiesemeyer, et al, "Astrophysical detection of the helium hydride ion HeH+," Nature, 2019

At this stage, the universe was not unlike a star, fueled by a hydrogen-helium fusion plasma. Photons could not escape due to the ultra-high temperatures and gravitation—known as the "decoupling era"—a time when the universe was so hot that electrons and could not remain bound to atoms.

The third element created during the Big Bang was lithium. Lithium is a metal and an ion, which means that it carries an electrical charge and makes a good conductor of both heat and electrical current.

Source: R. Krebs, "The History and Use of Our Earth's Chemical Elements: A Reference Guide," Greenwood Press, 2006.

Lithium is the lightest of all the metals. It is light enough to float on water. As it is a good heat conductor, it is often used in heat transfer processes. Its high electrochemical potential for ion storage makes it popular for auto, cell phone and computer batteries.

Sources: A. Helmenstine, Ph.D., "Lithium Facts: Li or Element 3," Thought Co. at thoughtco.com, 2019.

Also, lithium's nonlinearity makes it very useful in non-linear optics applications. It is used extensively in telecommunications products such as mobile phones and optical modulators, for such components as resonant crystals.

Source: "You've got the power: the evolution of batteries and the future of fuel cells," Toshiba, 2009.

As the "decoupling era" ended, hydrogen-helium gases expanded from the heat and pressures of the hydrogen-helium plasma. The wafer-light lithium was pushed above the expanding gases to form a wafer-thin metallic-membrane, like the skin of a balloon.

At the same time this lithium-membrane was forming, photons were finally released and able to travel at their maximum speed, 300,000 km/sec from the universe-core. Most of these photons were absorbed by the lithium-skin.

What did this merging of photons and lithium create? Remember, lithium is a natural ion and, as such, is an excellent conductor of electrons and electrical current. Along with the photons, the lithium-skin also absorbed and stored all the information the photons carried. The lithium-skin did not merely act as a universe-sized battery; it also acted as a semi-conductor.

Why does this matter? Let's take quick look at what semiconductors are:

What is a semiconductor?

According to semiconductors.org, semiconductors (also known as integrated circuits or microchips) are the "brains of modern electronics." As the name suggests, semiconductors are not pure conductors of electric current and they are not pure insulators. They do produce a variable degree of electrical resistance, depending upon their use and the amount of non-conducting metals used in their manufacture.

Semiconductors are essential components of electronic devices used in computing, communication, healthcare, military systems, and countless other applications.

"Today, there are more than 100 billion integrated circuits in daily use around the world—that's equal to the number of stars in our corner of the Milky Way galaxy."

Source: "What is a Semiconductor?" Semiconductor Industry Association at semiconductor.org. 2021.

The number of valence electrons (electrons in an atom's outer shell) determine the degree of conductivity a metal possesses. The metals most often used in semiconductor manufacture are silicon, germanium and gallium arsenide, which all possess at least four valence electrons. The more valence electrons a metal possesses, the quicker the electronic signal travels.

Source: "Semiconductor Materials," irds at ieee.org. (No date).

Semi-conductors receive, store, and transmit data using electrons. They are the "brains" of all computing devices using classical mechanical processes. The problem using computers with electron-semiconductors is that electrons are slower than photons. Photons travel at 300,000 km/sec. Electrons travel much slower.

"All current computer device technologies are indeed limited by the speed of electron motion. This limitation is rather fundamental, because the fastest possible speed for information transmission is of course, the speed of light, and the speed of an electron is already a substantial fraction of this."

Source: David DiVincenzo of IBM's Watson Research Center, "Computers are becoming faster, but their speed is still limited by the physical limitations of an electron moving through matter," Scientific American, Oct. 1999.

Can lithium act as a semiconductor?

In a scientific paper published by Optics Express in 2020, an international team of researchers have shown that wafer-thin lithium-photonic integrated circuits can provide "ultrahigh performance in electro-optic and nonlinear optical devices." They demonstrate that monolithic LN PICs fabricated on 4-and 6-inch wafers with deep ultraviolet lithography and show smooth and uniform etching:

"Thin film lithium niobate (LN) photonic integrated circuits (PICs) have recently emerged as a promising photonics platform for many emerging applications due to their superior electro-optic performance and large second-order optical nonlinearity…This can boost development in emerging large-scale PIC applications such as **quantum photonics and photonic neural networks."** (Emphasis added).

Source: "Wafer-scale low-loss lithium niobate photonic integrated circuits," K. Luke, P. Kharel, C. Remier, L. He, M. Loncar and M. Zhang, Optics Express, 2020.

Please note that the Luke research team is utilizing the close relationship between lithium and "photonics" (photons) to apply photonic integrated circuits to quantum mechanical applications and to photonic neural networks. Earlier chapters of this book and my previous book, ***Reincarnation: The Scientific Theory***, describe how photons play a critical role in how our brain's neural networks function and how the physiological process of individuality conservation (Reincarnation) occurs at death.

Quantum photonics, as mentioned in the scientific paper above, is a major component of quantum computing. What is so special about quantum computers?

Today's computers use electrons to compute data using 1s and 0s digitally. This is a classical mechanical operation much like using an electronic abacus. Quantum computers use "qubits" (quantum bits) instead of bits and take advantage of quantum mechanical superposition and photonic entanglement (remember Einstein's "spooky" paper we discussed earlier?) to make superfast and super-accurate calculations.

In 2019, the National Academy of Sciences published a research paper by Sara Gamble of the US Army Research Office. In her paper, Ms. Gamble is remarks:

"In reality, a quantum computer leverages entanglement between qubits and the probabilities associated with superpositions to carry out a series of operations (a quantum algorithm) such that certain probabilities are enhanced (i.e., those of the right answers) and other depressed, even to zero (i.e., those of the wrong answers)…

…The way quantum computers leverage probabilities and entanglement is what makes them so different from classical computers…Two of the more prominent areas (of use for quantum computers) are (1) quantum sensing and metrology (measurement), which leverage the extreme sensitivity of qubits to the environment to realize sensing beyond the classical shot noise limit, and (2) quantum networks and communications, which may lead to revolutionary ways to share information."

Source: S. Gamble, "Quantum Computing: What it is, Why We Want It, and How We're Trying to Get It," National Academy of Sciences, National Academy of Engineering, Frontiers of Engineering: Reports on Leading-Edge Engineering from the 2018 Symposium. Washington, D. C., National Academies Press, 2019.

We see from the *Optics Express* and *National Academy of Science* papers, above, how researchers are slowly discovering the relationship between photons, lithium and quantum computing. At some point, they will discover the special relationship between photons, lithium processing and memory storage. Once that happens, expect a giant leap in technological innovation that will exceed anything we have experienced to date.

One example of the technological fruit that will grow from these discoveries is artificial intelligence. Once photons are "married" to lithium quantum computing processers with entanglement, artificial intelligence will become self-aware.

Won't that be a "face in the pond" staring-back-at-us moment! Better get ready for a self-aware artificial intelligence, because it is just a matter of time before it is "born."

In her paper, Gamble mentions that developing quantum computing has been a struggle due to the difficulty involved, with our limited technology, to prolong quantum states necessary for the quantum computing to process its qubits.

"First, qubits need to be protected from the environment because it can destroy the delicate quantum states needed for computation. The longer a qubit survives in its desired state the longer its **'coherence time.'** From this perspective, isolation is prized. Second, however, for algorithm execution qubits need to be entangled, shuffled around physical architectures, and controllable on demand."

(Gamble, cited above).

What you are observing from these research papers is humanity's first steps toward mimicking how the early universe employed quantum entangled computing to become "coherent" on its own.

The difficulty Gamble alluded to regarding quantum-state stability is the interface between the photons making up the qubits and the "architecture" of the semiconductor itself. Researchers are having difficulty finding a metallic-molecule that is harmonious with quantum states necessary for superposition-stability but also practical enough to serve as a semiconductor. Just a suggestion, but perhaps they should try a lithium-based molecule for its quantum computer superstructure. What do you think?

The universe did. How did the universe become self-aware? No one knows for sure. In fact, I do not believe anyone has attempted to theorize such an event—until now.

The time has come to make the attempt. The Universal Mind, like the universe and all living organisms within the universe are systems. They function as systems. Like any other system, the Universal Mind evolved incrementally:

1. Following the "decoupling era," photons (carrying all the information defining the universe) sped out from the center of the universe at speeds much greater than the universe, itself, was expanding. As a result, these photons collided with the lithium-skin at the universe's edge.

2. Over time, these photons covered the lithium-skin with a quantum mechanical photon-shell.

3. These photons began sharing information with each other through ion-exchange. and quantum entanglement.

4. At some point in time thereafter, these photons became quantum-entangled; a quantum mechanical level of existence akin to infinite union over infinite distances.

5. The now quantum-entangle photons of the lithium-skin began an exponential evolution of consciousness from individual, information-carrying particles; to inter-photon communication; then to individual-photon decision-making; and finally to a unity among all photons in the lithium-skin.

6. Self-awareness was inevitable.

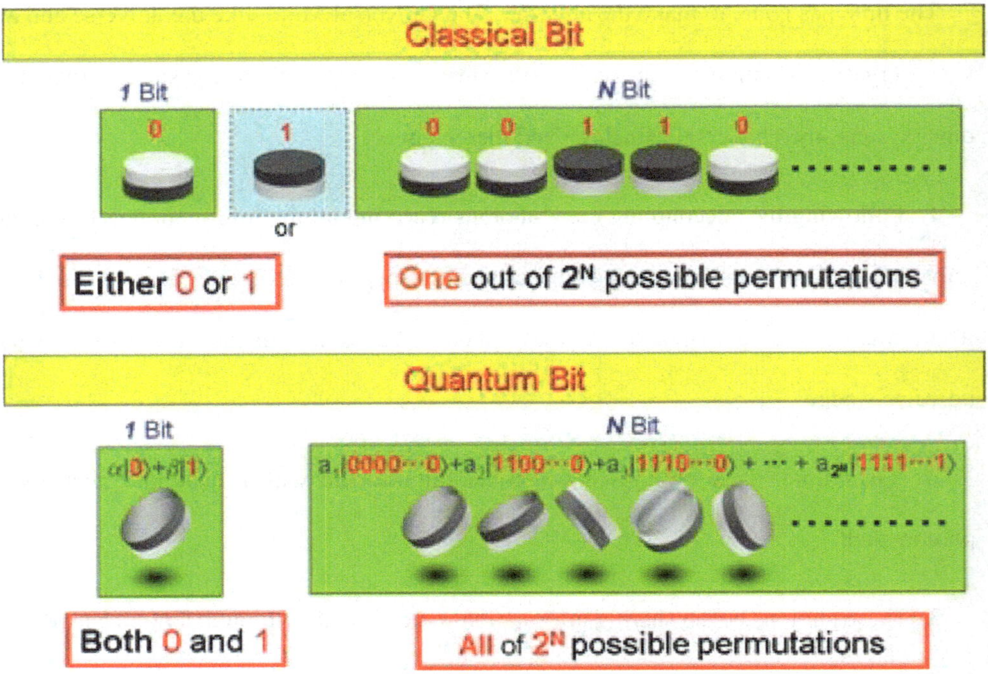

Illustration No. 19. Classical and quantum bits. copyright 2022

Illustration No. 19 demonstrates how differently the typical digital computer and a quantum computer function. A "bit" is a single unit of information using electricity (0 for off, 1 for on). Computing using 1s and 0s is called "digital computation." All computer programming today uses some combination of 0s and 1s in their algorithms.

Quantum computers use "qubits" (Qbits) of photons instead of "bits" of electrons used in ordinary computers and electronic devices. In quantum computing, a "qubit" or quantum-bit is the basic unit of quantum data. A qubit is a two-state quantum mechanical system where the photons exist in superposition (both as 1s and 0s simultaneously) with each other.

Source. quantumcomputing.com

We can now see how, during the early universe, lithium would have been a perfect semiconductor for photonic wave energy. The photon-carrying, lithium-skin enveloping the early universe would have served as a giant quantum computer, capable of coherence and ultimately, self-awareness.

But what does it mean to be "self-aware?" Does every human have the capacity for self-awareness? Are people who refuse to be injected with medically-approved anti-Covid vaccines, yet rush to the closest veterinary clinic to buy Ivermectin (an anti-worm medication for horses), self-aware? Are people who take flu shots without questioning them, but who stubbornly refuse to be vaccinated against Covid self-aware? How self-aware were the "patriots" who allegedly participated in the fatal riot and attempted insurrection by storming the national Capitol on January 6, 2021?

The point is that "self-awareness," like everything else in our quantum mechanical universe is—relative. Consciousness does not equate, or even require, self-awareness. All life has the intrinsic ability to be conscious. But all that means is that the brain is receptive to sensory inputs and the brain acts upon those inputs to create an intentional response.

Is this, and this alone, sufficient for self-awareness?

Physicist and noted author, Dr. Amit Goswami, Ph.D., explored the concept of a "self-aware" universe decades ago. He recognized then that the human brain has the ability to comprehend events and share communication with others that could only happen if the brain was somehow connected to a "nonlocalized" consciousness."

"Could it be that a quantum mechanism in our brain, operating in ways similar to the laser, opens itself to the supervention of nonlocal consciousness, with the classical parts of the brain performing the role of measuring apparatus for amplifying and making (if only temporary) records? I am convinced that the answer is yes."

Source: <u>The Self-Aware Universe: how consciousness creates the material world,</u> Dr. Amit Goswami, Ph.D., (with Richard Reed and Maggie Goswami), Penquin-Putnam, New York, 1993.

Dr. Goswami's book proposed a new way of looking at the brain-mind connection. It theorized this connection formed a link between the brain's physiological matrix as sensory analyzer and measurement device and what physicists describe as a "nonlocalized" mind—capable of communication with a "universal consciousness" linking all life.

I share Dr. Goswami's vision of a universal consciousness linked to each human brain by a nonlocalized, quantum connection. But we must take his vision two steps further.

First, there would be no quantum connection with the Universal Mind without photons. It is the photon that creates quantum mechanics possible. Without photons there is no quantum entanglement and no Universal Mind.

Second, Dr. Goswami did not realize that this quantum connection makes Reincarnation a consequential, causal result of the Big Bang and the natural laws derived from it. But we know from earlier chapters in this book that the Dalai Lama understands the connection, don't we?

Likely, so do hundreds of American college physics and philosophy professors. But they are too intimidated by Republican-fascist attacks on academic freedoms to acknowledge it. They have worked too hard, for too long, to risk their status and their pensions to risk rocking the ideological "boat."

CHAPTER 14

Universal Reincarnation and Infinity

Let's make some hot tea, shall we?

Everyone knows how to make hot tea. You put water in a tea kettle and place it on the stove and heat it up.

But how do you know when the water is hot enough to make tea?

You hear the tea kettle whistling, that's how. Heat makes the water molecules move faster and faster until there is a "change of state" of the water from liquid to a gas. Because the kettle is a confined space, the gas is under pressure and seeks lower density—the hole in the spout. Hot, pressurized water vapor pours out of the spout. You hear a whistle.

Close the hole in the spout and you have a reasonable approximation of the early universe during the plasma-state known as the "decoupling era" that we discussed earlier. The universe, like a tea kettle with no spout, is a closed-system.

Nothing enters that wasn't there to begin with; nothing leaves. The system is closed. The molecules may move faster and faster, becoming ever more random in their behavior. The increased randomness increases the intensity, the density of the steam being released.

Why do we care about a stupid tea kettle? We care because a tea kettle filled with water under heat offers us a reasonable illustration how the second law of thermodynamics works. In any closed system, the longer the system functions, the more random its behavior becomes.

Why?

The longer a closed system operates, the more information is created that cannot be destroyed. Therefore, every action creates a reaction that is necessarily more random and less predictable than the one before it, which results in its own causal response, and so on…to infinity.

Now, close the spout on the kettle. Don't use your finger. You'll get a nasty steam burn. Once you've sealed the spout, what happens inside?

Instead of pouring out of the kettle through the spout, the steam will circulate around the edges of the kettle in a hot, vapor cloud. Some molecules of water vapor will fall back into the heated water below. Other molecules, however, will condense upon the stainless-steel inside of the kettle.

That whistle you hear when steam leaves the kettle, do you know what that is? That is information. It is telling you the water is boiling hot and ready to make tea.

Notice anything you remember from an earlier chapter? What we have just described is what likely happened during the "decoupling era" as photons were trapped within the plasma-field and became embedded within the lithium layer coating the early universe of hydrogen, helium and lithium.

The "bubble" skin of the universe consisted of a layer of photon-absorbed lithium. The hydrogen-helium gas within the "bubble" were the molecular building blocks of all the matter that now exists within the universe (except for dark matter, the make-up of which is yet unknown).

Reincarnation is simply a causal-function of the Second Law of Thermodynamics which states that uncertainty (randomness) always increases with time in any closed system (such as the universe). This law has two consequences:

1. Time is a relativistic phenomenon within any closed system. No particle within a system can go back in time within that system. Whatever a particle does within a system adds information (randomness) to that system. Taking that particle back in time would, in effect, delete or erase the information the particle created in its movement within the system—violating the Second Law of Thermodynamics which states that information cannot be lost.

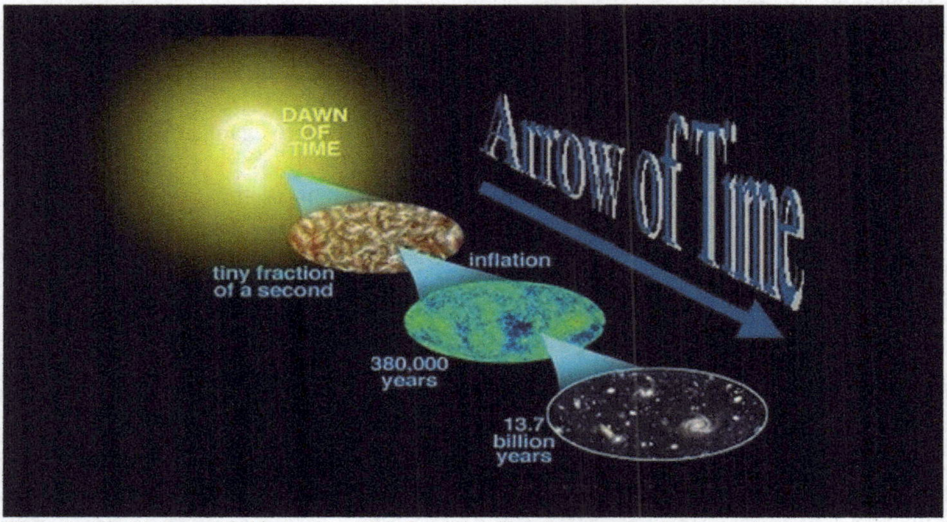

Illustration No. 20. A modified image from WMAP website. By unknown author licensed under CC.

Illustration No. 20, above, depicts how the universe expanded over time. Notice how the universe expanded from a concentrated plasma-field where all matter and energy were bound together; to the universe of today, filled with galaxies, photons, gas cloud filaments, dark matter and dark energy. Note that the "Arrow of Time" incorporates two concepts in one—the "arrow" is both a state-of-being (Time) and information. Existence cannot occur without information being created along with it.

Remember, the First Law of Thermodynamics states that in a closed system, energy can never be created nor destroyed. It may change state from energy to matter and back again, but the total amount remains stable.

As the universe experienced each stage of its evolution, it changed states, adding to the information of its existence and its relative randomness. The illustration above is also an example of how a closed system like the universe experiences ever-increasing entropy or randomness over time.

Randomness or entropy is just another aspect of information. The more a system changes, the more random its particles become, and the more change has occurred withing that system. The change experienced by the system is the same as information. Information is, after all, merely a measurement of the state of system as it changes over time.

Remember, the Second Law of Thermodynamics states that in any closed system like the universe, as time passes, entropy or randomness of the matter within the system increases.

Some describe this law as the "nondeleting" and "noncloning" principles of the Second Law. Entropy or information cannot be lost or "deleted"—nor can it be added to or "cloned." Therefore, traveling back in time would delete the information already created, which violates natural laws, and therefore, cannot occur.

Individuality creates a state-of-change. Therefore, individuality becomes a system just like the universe. As we exist in time, we add experience to our consciousness in the form of stored memories. These memories are used by our brain to create our unique individualities. Each individuality is unique from every other. Therefore, individuality is like a miniature universe where each new memory creates a corresponding change-in-state within our unique individualities, which equates to any change-of-state experienced by the universe.

Reincarnation is the reverse or mirror image of this "non-deleting" principle. Delete any of the change-of-states the universe has evolved through and you delete the universe itself. The Second Law of Thermodynamics makes this deletion impossible, just as it is impossible to travel back in time within one's own frame of reference.

The same principle applies to human existence because our consciousness shares the same closed system as does the universe. As self-aware systems, we, too, create changes-of-state which—if any were deleted or lost—would result in our never having existed in the first place. Delete any of the memories that form your unique individuality, and you delete yourself.

But never fear, because such an outcome cannot happen as it would violate the Second Law of Thermodynamics. Time is a river and information is the sediment carried along with it, altering the bottom forever as it flows. There is no backwash. Time and information can only move forward together.

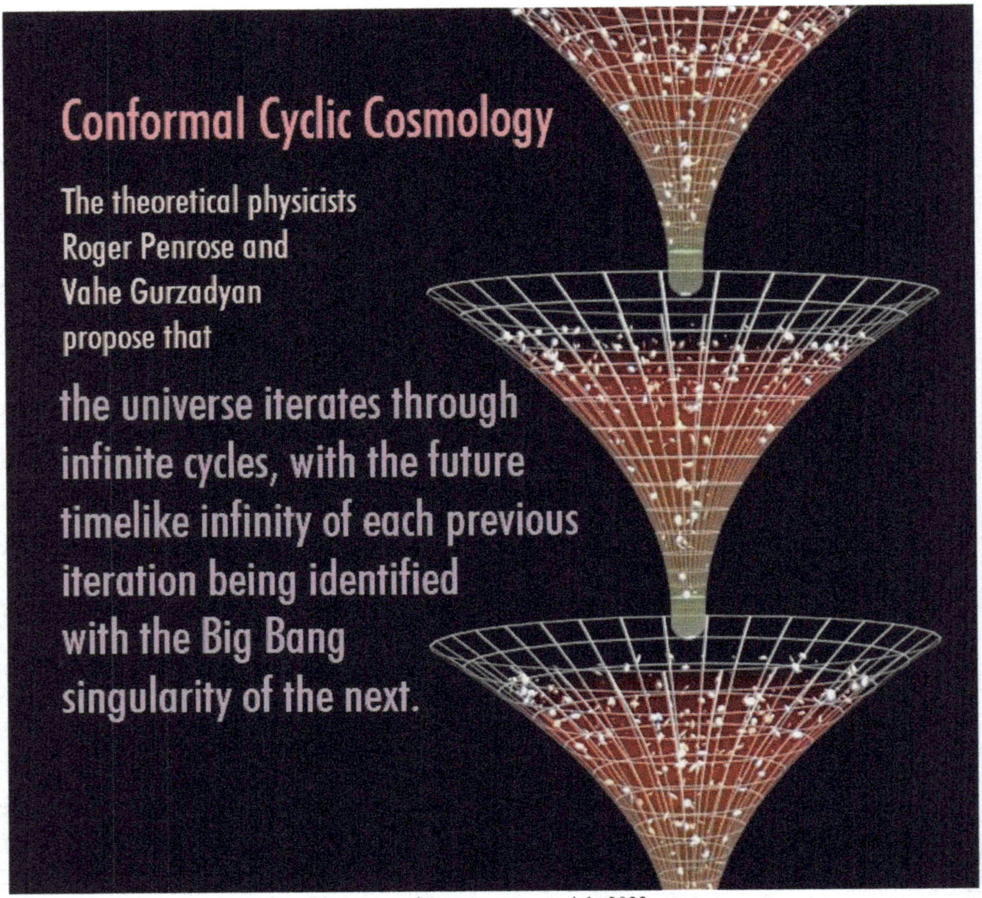

Illustration No. 21. Penrose redux. Pinterest at pinterest.com. copyright 2022

Above, in Illustration No. 21, we return to Professor Penrose's Conformist Cyclic Cosmology to consider the role death plays in the process of conserving information—or in the case of us humans, conserving our unique individuality.

What would happen if Reincarnation or rebirth did not exist? This is like asking what would happen if the Second Law of Thermodynamics no longer applied to information?

The answer, although complex, is inevitable. Time is, in itself, a "change-of-state" forcing anything within its frame of reference to change with it. Any change-of-state within a system creates information. Therefore, Time is information. Anyone within the frame of reference of a system cannot alter the information within that system.

Time cannot be cloned or deleted from within its own frame of reference. We cannot add Time to Time any more than we can delete Time from Time. This is so, because everything that exists within the universe exists within the frame of reference of universal change-of-states—which includes Time.

This is why the Second Law of Thermodynamics—a conservation law which originally was thought only to apply to physical changes-of-state (steam turbines, tea kettles, etc.)—has since been linked to information conservation and our own existence.

We humans cause change. We are change-of-state actors, as it were. But we are more than that. We are living information. Our individualities are unique data-sets which cannot be duplicated anywhere else in the universe.

What created evolution? We didn't. Evolution created us. But evolution did not create life—it merely designs life. Life created evolution. What created life?

The Universal Mind created life. This is the only logical explanation. Nothing else existed to accomplish this. The only other possible explanation for life to occur is random chance. Self-aware photons interconnected into a lithium-bound web of electromagnetic intelligence that exists throughout the universe.

If you believe life was the result of random molecules, then you fail to take into account universal inevitability, which we have observe as the "Johnny Appleseed effect."

John Chapman (aka "Johnny Appleseed") was born in Leominster, Massachusetts in 1774. His claim to fame was that he spent much of his life traveling the then Western frontier of western Pennsylvania, Ohio, Indiana and Illinois planting apple seeds for apple cider wherever he went.

Source: N. Gelling, "The Real Johnny Appleseed Brought Apples and Booze to the American Frontier," Smithsonian Magazine, 2014.

The point is that if you throw out enough apple seeds, the growth of apple orchards no longer can be thought of as a random or uncertain result. Throw out enough seeds and apple trees and their fruit become inevitable.

On a universal scale, galaxies are the forests and planets are the trees. We humans are one variety of life in a forest of life. Each life-form arose as did every other—from the photons created during the Big Bang.

Every species shares the same building blocks of energy. We may appear different to each other due to evolutionary variants caused by the physical environments of our planets of origin, but fundamentally, we are all the same. We all share the same conscious link to the Universal Mind—and we all share the same Reincarnation effect.

CHAPTER 15

Quantum "nonlocality" and Quantum entanglement

We're going to discuss some serious physics here but don't be scared. We're going to do it in a way you can understand. So, relax and work with me on this.

Photons can do amazing things. Two properties photons have are quantum entanglement and quantum nonlocality. We've discussed these concepts before, but now we need to discuss them in more detail because it's time to take our next BIG step. We are about to learn how photons create consciousness and, therefore, the Universal Mind.

What are these two properties? They are not the same. Quantum nonlocality is NOT the same as quantum entanglement. Quantum entanglement is the cause of quantum nonlocality.

Quantum entanglement is a quantum mechanical process in which two or more photons become linked with one another. Scientists have only recently discovered this is possible and know of only one way it can be done. So far, the only way we can entangle photons is by forcing photons together using very strong magnetic fields. While it is certain that this is not the only or most elegant means of entangling photons, it is all we know at this time.

Einstein's description of photon "spooky behavior at a distance" suggested the existence of photon-entanglement and nonlocal existence back in 1935 when he and his co-authors wrote that photons produce "spooky action at a distance."

Source: Einstein, A., Podolsky, B., and Rosen, N., "Can quantum-mechanical description of physical reality be considered complete?", Physics Review, 777, 1935, retrieved at journals.aps.org.

Consider the implications of Einstein's "spooky action"...

"Nonlocality describes the apparent ability of objects to instantaneously know about each other's state, even when separated by large distances (potentially even billions of light years), almost as if the universe at large instantaneously arranges its particles in anticipation of future events.

Thus, the quantum world, despite what Einstein had established about the speed of light being the maximum speed for anything in the universe, instantaneous transfer of information does appear to be possible...Nonlocality suggests that the universe is in fact profoundly different from our habitual understanding of it, and that the 'separate' parts of the universe are actually potentially connected in an intimate and immediate way."

Source: "Nonlocality and entanglement," Physics of the Universe at physicsoftheuniverse.com.

Einstein's theory of photon "spooky action at a distance" has been tested and proven repeatedly in a series of recent independent experiments. In 1998, Professors Furusawa and Sorenson reported the first successful experimental quantum teleportation of photon energy. In 2016, scientists successfully teleported entangled photons over a 30-kilometer optical fiber quantum network. Applications of quantum entangled photons are already on-going in the communications and computer fields.

Separate and independent from the classical mechanical functioning of the brain is a background awareness that is "nonlocalized." Our scientific instruments detect it as part of the cosmic microwave background (CMB) radiation. This ability of photons to exhibit "spooky action at a distance" must be a function of this photonic wave "nonlocality."

Sources: A. Furusawa and J.L. Sorenson, et al, "Unconditional quantum teleportation," Science, 1998, retrieved from sciencemag.org.; Qi-chao Sun and Ya-Li Mao, "Quantum teleportation with independent sources and prior entanglement distribution over a network," Nature Photonics, 2016; and R. Valivarthi and M.G. Puigibert, et al, "Quantum teleportation across a metropolitan fibre network," Nature Photonics, 2016.

Quantum entanglement is no longer just a theory. Quantum teleportation and other forms of nonlocality are no longer mere theories. They are no longer science fiction. They are science fact.

We absorb photons from both the Big Bang as well as those emitted from our sun and other stars throughout the universe. We also continuously emit photons from our own brains as the result of biochemical reactions and ion exchanges.

We are a part of this Universal Mind. Our brain functions interact with entangled photons exhibiting nonlocality capabilities all the time. This entanglement gives each of the entangled photons a shared and instantaneous knowledge of all the information the other possesses. Our brains also absorb and emit these same entangled photons.

We have no alternative but to accept as a fact that our minds are in direct contact with the Universal Mind through these entangled photons. While our minds are always connected to the Universal Mind, we are not consciously aware of it. When we do experience conscious contact with the Universal Mind, our consciousness is raised exponentially. This is called "superawareness."

This "superawareness" is a heightened sense of existence; one in which we become increasingly aware of the infinite information within the Universal Mind. The more one's consciousness is enlightened or raised in awareness, the more knowledge can be shared with the Universal Mind and the longer our limited brains can retain it.

You do not need to be religious at all to share in this communion. All it takes is being receptive enough to "tune in" to its presence. It is this "superawareness" that humanity has confused with "spirit" or "God." This "superawareness" is what we feel when in conscious contact with the Universal Mind.

Can we communicate with the Universal Mind? Of course, we can. All it takes is stillness of self—the type of stillness which requires enlightened self-knowledge. Regular meditation is very helpful in achieving this enlightened state of mind. No matter our religion or philosophy, everyone benefits from regular meditation.

You want to touch the Universal Mind and be consciously aware of it? Of course, you do. The Universal Mind is like a priceless gem, but it is buried deed, very deep underground. To reach this gem, you have to dig…and dig…and dig…and dig some more.

How long do you have to dig? I can't answer that. Everyone is different. Some have to dig longer than others. It just depends upon how enlightened your past lives have made you.

What kind of shovel do you use? You do know this gem-story is just an analogy, right? The "gem" is the Universal Mind. The "shovel" is meditation and living an ethical and honest life. Sorry, Donald Trump. Sorry, Ronny DeSantis. As long as you both lie and use hate, prejudice and intolerance as weapons, meditation will do you no good. As long as your personal lives remain evil, meditation does you no good.

It is not my intention to turn this book into a meditation text. However, as we are discussing contact with the Universal Mind, it is necessary to explore some very basic principles.

First, meditation is a very personal experience. I am an expert only of my own experience. Everyone can meditate. You can become an expert of your own meditation experience without spending much money. All it takes is you making the personal commitment to do it.

The Internet and media provide an unlimited source of free knowledge on meditation. However, too much information is often overwhelming for a novice and can do more harm than good.

Your best and most trusted mentor is you and your inner self. Think of the concept of investigating meditation as you would think of walking a trail in the forest. Search out your local bookstore or public library for relatively inexpensive books and media that give you a variety of meditation techniques to learn from. Go with what you feel comfortable with. Then adjust your technique to suit your own individual journey as your own consciousness guides you.

If you choose to have a personal meditation instructor, investigate several before settling upon one whom you enjoy meditating with and whom you trust. Beware anyone who appears preoccupied with taking your money.

Meditation training is a trust and should never involve more than a nominal fee for expenses. Meditation is not ball-room dancing. Meditation is not televangelism. It should not be a for-profit-business. The goal in leading a meditation group should not be to get rich. If you get the feeling that your meditation leader is hungrier for dollars than for your spiritual growth, it's time to find another instructor.

Your instructor should not be rocking expensive jewelry, rich-Italian leather loafers, driving a brand-new Bentley and bragging about his condo in Aspen or her beach house in Maui. The very image of material excess surrounding one claiming to be committed to a "non-material" life-style such as meditation should be an instantaneous revelation that your "instructor" is more interested in your finances than your mind or soul. These phonies are to be avoided.

Avoid those who claim it will cost a lot of money to get you from where you are to where you want to be. Avoid the con-artists and hucksters who rake in money as televangelist-fake-Christians selling you TV-salvation like it's an opal necklace on some shopping network.

Meditation only takes as much or as little time as you want to spend doing it. You are never too young, too old, too rich, too poor, or too sick to meditate. You can sit in a chair or on the floor. You can lay on a couch, the floor, or a bed.

Do not force meditation upon yourself when you are too busy or have too much to do. These concerns will only interfere with your ability to meditate at that time. Get your work done first. When your work is finished, you will be in a much better frame of mind to relax and meditate.

I've meditated during lunch hours or in a chair after coming home from work. If you fall asleep during this meditation, do not be discouraged. All it means is that your body needed rest more than your mind needed its meditation for the time being. Rest your body first, then return to meditation at the next convenient moment.

People confuse the purpose of meditation as a process, when it is a state of mind that we are seeking. Do not get hung up on whether your process of meditation is "ok." There are many ways to meditate. The point is that there is no one way, no single process for "proper" meditation. Find what works for you and grow with it.

Over the years, I have mainly used two different meditation techniques. I cannot say they will work for you. But they have for me.

The first method is what could be called "focused" meditation. This method can be used with many different formats. For example, as a child I spent hours and hours every day fighting for each breath. It was hell—and it never ended. To cope mentally, I instinctively taught myself distraction-training.

While I was suffering from a bad asthma attack I could only breathe by physically forcing my chest up to fight for air. This wore me out physically and mentally as well.

At four-years-old, I didn't have much energy to begin with and there were no medicines back in the 1950's to help. Whether I survived or not was up to me and how stubbornly I wanted to breathe. I lived from breath to breath.

This took its toll on me. I almost died several times. Through these experiences I learned that what I needed was a "free-space" within my consciousness with which to be free of the daily torture my body was suffering.

To create a "free-space" within my mind, I had my mom tickle my back, arms, or scalp just to give me a physical sensation to focus upon something other than the "elephant" sitting on my chest. The conflicting physical sensations this created had a curious effect upon my consciousness. The conflicting physical sensations separated my mind from my physical surroundings.

My inner "self" became an empty space existing between two words—"pain" and "pleasure." Painful sensations from asthma had to compete with the pleasurable sensations of skin-tickling for attention within my consciousness. After a time, the two cancelled each other out. These two competing and opposing sensations were both replaced with space or emptiness. As a child, I just called it bliss.

This became my childhood routine. Every night I cajoled my poor mother to tickle me for hours and hours. Despite losing the sleep she needed to go to work the next day, she never refused. She was my life-line. I would not be here today without her sacrifice.

Focused meditation is used by many. Some place a rose or other flower in front of them and gaze into it as a distraction from the material world. Many rely on "breathing focus;" which is concentrating on nothing but one's own breath. Others use music, repetitive sounds, or art as their focus.

Each of these methods uses the same principle of "conflicting stimuli" I discovered as a child. Use the music or your breath as a conflicting signal to compete with the busy-ness your mind is running on. This busy-ness is like the white-frothy foam that leaps up off the sea as the waves crash upon the beach. To submerge your being in the Universal "sea" you must ignore your mind's superficial "foam."

Another form of meditation I have relied upon is what I call "unfocused" meditation. In this method, I simply allow my mind to wander freely, wherever it wants to go.

At first, your mind may fill with the random daily thoughts and emotions that have become attached to your surface consciousness like dirt on a shirt. To allow your deeper consciousness to come forward, allow those superficial thoughts and emotions to float away like a laundry wash cycle.

The main thing is not to force this daily mental debris away. It will seem counterintuitive at first, but believe me, any effort expended attempting to direct the mind away from where it wants to go is counterproductive.

Let these surface thoughts float away on their own. The less effort you apply, the quicker your mind will begin to sink. Image what is like to float on the water in a pool. Forcing superficial thoughts away is like kicking with your arms and legs while you try to sink into the water. It is counterproductive. All that kicking forces your mind up toward the surface when you really want your mind to sink.

Trust your mind. Just relax the mind and let it take over. Just as all water seeks its own level, the mind will always seek out its "home"—the Universal Mind. Your mind knows instinctively where its "home" is and will seek it out automatically without any effort from you. All you need is patience and the willingness to let go.

Use your common sense. Do not try to meditate while driving a motor vehicle or when you are involved in any physical activity or manual labor that could be physically dangerous to yourself or others. Meditation tends to relax you to the point where you can often fall asleep without trying. Find a safe and quiet place where you can relax before meditation.

Meditation is not for the impatient. Meditation is not for those who seek wealth or fame. Meditation is a life-long journey of turning away from all these forms of self-worship. You will have successes and failures. But the longer you meditate (years), the closer you will come to your goal.

The goal of meditation is increasing communion with the Universal Mind. But to do so you've got to meditate.

You can meditate alone…

Illustration No. 22. Meditating alone. copyright 2022

You can meditate with a group…

Illustration No. 23. Meditating with a group. copyright 2022

You can meditate in a chair…

This Photo by Unknown Author is licensed under CC BY-NC-ND

Illustration No. 24. Meditating in a chair. No disability can prevent us from meditating. copyright 2022

You can meditate anywhere…

This Photo by Unknown Author is licensed under CC BY-SA

Illustration No. 25. Astronaut floating in space. copyright 2022

Meditation is an internal reality controlled by the link existing connecting the deep recesses of your consciousness with the Universal Mind.

We seek reunion with the Universal Mind from which we all originated. But because this urge is rooted so deeply within our individual and collective psyche that its motivational force acts independent of our conscious mind.

While our consciousness is distracted by the sensory stimuli of each day's events and the emotions they create, this deep urge for communion operates covertly in the background like a computer algorithm in a hidden data file. Along with our evolution-created herd mentality, this may explain why humans so easily fall prey to religious scams.

No matter how obvious a scam may appear to most people, there will always be some who need to be a part of something. The majority would rather join a group simply for camaraderie and a shared sense of togetherness than exist alone—even if that group is about fake, contrived conspiracy theories or a harmful cult.

But true communion with the Universal Mind requires just the opposite of camaraderie. Communion with the Universal Mind requires solitude and mental withdrawal from the rest of the physical world. This can only be done through meditation.

How can meditation help you connect with the Universal Mind? The benefits of meditation begin with you learning about yourself and how to turn your "active" mind off. Your "active" mind is the part of you that handles everything you do on a day-to-day, routine basis. It is your "active" mind that keeps you up at night after another stressful day at work.

Meditation can help your "active" mind wind down at night without the need for alcohol or drugs. But meditation is not prayer. People "pray" because they want their god to do something for them. Prayer is about desire and need. Prayer is the glorification of, and submission to, the self.

Meditation is the exact opposite. Meditation is the search for "selflessness" through self-denial, the opposite of self-glorification. You can tell bullies and tyrants like Trump and DeSantis don't meditate sincerely. If they did, they would NOT have become who and what they are.

Meditation forms a channel through your "active" mind into the deeper core of your consciousness. The more you meditate the quieter your "active" mind becomes and the deeper you reach into your consciousness. With continued meditation over time, you begin to experience a sense of higher reality.

What is that "higher reality?"

Is it really possible we can be in conscious contact with the Universal Mind? Ever since that night in the Arizona desert, I have wrestled with what or who I was in telepathic contact with that evening. The "voice" that told me about life on other worlds throughout the universe being like grains of sand in the desert.

Was that "voice" in my head an alien species or was it the Universal Mind? Perhaps you remain unconvinced of either one. There have been other instances where something "spooky" happened to me that defies classical mechanical explanation.

As I grew into my teens, I began having "waking dreams" which put my awareness inside the mind of others experiencing their lives. For many years, these dreams occurred twice or three times a week. They still plague me, but not as much.

These dreams are not a pleasant experience. Over ninety per cent of these dreams involve violent events and very negative emotions. Strangely, almost all of these dreams are NOT about me. They involve others, total strangers to me. I have never met them.

Then, months later, I would see on television or hear over the radio some incident that was exactly like the dream I had experienced months before. I kept these to myself—until now.

Causality is not defined by space/time. Causality is independent of both. In fact, the Universal Mind may be a causality machine creating progression-chains of thought possibilities which act as "thought DNA," communicated to us via the very photonic waves the Universal Mind and we share. I decline to use the term "we both share" because this phrase inaccurately describes our relationship with the Universal Mind. "Separate, we are not," Yoda remarked wryly.

Where or when something horrific or wonderful will occur in the future is indeterminate. What does become certain, given the proper causality, is that these events WILL occur.

Causality is no more than temporal inertia—like evolution. Begin a walk and eventually you arrive at your destination. You may not have been aware of where you were headed, but some part of your consciousness led you on until you reached your destination. You can consider that a weekend trek into the forest or your life. It is all the same.

While the Universal Mind acts as a guide to moderate or influence our thoughts and, thereby, have an indirect influence upon our choices; those choices remain our own. For better or worse, we enjoy the illusion of free will.

But the Universal Mind cannot keep a drunk driver from killing you. It may influence him or her not to drink and drive to begin with. Or it may influence you, albeit indirectly, not to go out that night, or to take a different route.

But once you bust open the envelope, the laws of physics cannot be undone just to save your ass. But if you remain within the envelope of physical laws, the Universal Mind can reach out and change your life.

I attended law school at George Washington University law school in Washington, D.C. I thought I could get an apartment the week before classes started.

That was stupid. There were no vacant apartments anywhere. I looked in Arlington, Alexandria, Georgetown, Dupont Circle without success. When classes began, I became desperate. I began living in the International Youth Hostel in downtown D.C. For five bucks a night, I shared a room with three techno-pop dudes from Germany and slept on a bunk-bed.

Sorry, forget the sleep part. The three techno-poppers from Germany stayed up all night, every night. Two o'clock in the morning their boombox was blasting electronic noise at 120 decibels, while the techno-geeks were screaming in German. I couldn't tell if their screams were lyrics or just insane rants.

I never did learn their names. Instead of the sleep I needed to be fresh for law school classes, I got six to eight hours a night of their jumping up and down in the room to their techno-pop at max volume, while waving brightly-colored glow-sticks and light wands around in the dark.

I couldn't sleep. I couldn't study. By the second night, I thought I had gone mad.

To make matters even worse, I was facing a critical decision. As the second week began, a deadline loomed ahead of me which threatened to end my law school career before it really even started.

Law school is very expensive, even if one attends a state supported university. I had applied to several public law schools in the event I was not accepted by the law school I really wanted to attend—George Washington University. Being a private university, GWU was much more expensive.

I had borrowed a considerable sum with interest from a savings and loan for tuition. Even in 1980 money value, the cost of a law school degree was not cheap--$8000 per semester for tuition alone. Living expenses, fees, books, and whatnot were extra.

But all these financial requirements were irrelevant unless I found a solution to my over-riding dilemma. Living in a dorm-like hotel with "wild and crazy guys" from Europe was a deal-breaker for me. If I couldn't study or get enough sleep, I was going to flunk out. I needed a permanent place to live for the next three years—and I needed it immediately.

GWU only gave full refunds up to the end of the second week of classes. That made my decision easy. If I couldn't find a permanent place to live before the refund deadline, I would formally withdraw from GWU Law. That would be that.

That first week of law school was the most difficult week of my life. First semester law school at GWU was a Navy SEAL "Hell Week" for the mind. The SEALs only have to endure "Hell" for one week. Law school has a traditional "Hell" they put students through. It's called First Semester. The object is to make you quit and the professors áre very good at finding the right psychological "buttons" to push each first-year student to the edge.

Our torts professor could have been the inspiration for *Harry Potter's* "Professor Snape." He was a brilliantly successful personal injury trial lawyer who knew how to get students to crack. One day, during the second week of classes, randomly chose a particular young female student to stand up in front of the class and read her outline of one of the dozen cases we were assigned to read the night before. Her soft voice could barely be heard as she read her outline (called a "brief").

"Now, Miss Jones," bellowed the professor. "Explain to me what the court would rule if the facts were as follows…(he read off ten facts so fast no one would have been able to recall them all) and why."

A class of a hundred students sat expectantly waiting for her reply (believe me, no one wanted to be called on by this professor—ever). There was no answer. She stood as though frozen to the spot.

Ten seconds passed by…nothing.

Twenty second passed by….still nothing. At this point, some of the morons in the back row began laughing at her. The professor did nothing to stop them.

Thirty seconds passed…The young woman had begun sobbing. Cries of "Let her sit" from around the lecture hall replaced the mocking laughter from the back row.

"Ms. Jones!" The professor did nothing to hide the derisive sarcasm. "You may sit down now. And thank you so very much for your thoughtful response. You obviously have a future as a paralegal."

Ms. Jones fled from the lecture hall, sobbing as she ran. None of us ever saw her again. She likely dropped out of school that day. But she was just one of his victims. Hardly a day passed when someone did not drop out of school. I vowed I would not become one of them.

But I was living two-lives. During the day, I pretended everything was normal. I got breakfast and morning coffee at the student center and went to class like everyone else. During the day, I forgot about the crazy Germans. I forgot about the screams, sirens, and other street noise that inundated the International Youth Hostel in downtown D.C.

But as soon as classes ended, I had a hell-hole for a place to live and no chance to study. And time was running out.

The second weekend began with hope. Maybe I would find a place to live. I looked all over. I went to every rooming house, boarding house, apartment building and rental house I could find in the D.C. area.

The second weekend ended just as it started—no luck. I was no closer to finding a place to live than I was when I arrived in Washington weeks before.

That Sunday night, I gave myself the bad news. I would keep trying through Wednesday night. But, if I didn't find a place to live by then, I would withdraw from law school the next day. That would be the day before the deadline for refunding me my borrowed tuition.

By now, I had just about given up. I felt I had tried every possible option. There just was nothing available. I knew I should have secured a place to live over the summer, but I had been stupidly naïve.

It was a horrible feeling. This was a failure of majestic proportions. I was the first of my family to even go to college. Now my parents were expecting me to become a lawyer. By waiting to the last minute to find a place to live, I not only let myself down—I had failed my parents. I felt like an idiot. It was humiliating.

Tuesday night at the Youth Hostel was awful. The techno-pop boys were friendly but totally oblivious to my law school "death march." That Tuesday night I pulled my pillow over my head to drown out their noise. I prayed...hard.

"Whatever you are out there. Hear me, please. You wanted me to be a lawyer. I am here because you told me to be here. I am here and I am failing. I have lost hope. If you want me to be a lawyer, please give me a dream tonight where I can find a place to live by tomorrow. If you cannot, I will have to withdraw. I have failed you. I am sorry."

Going to sleep that night was liberating. My two weeks of hell were over. The worst that would happen would be that I end up back home Thursday night, got a good meal and a decent night's sleep for a change.

Then…something wonderful…

I had a "waking dream." The dream begins with me standing at the corner of 21st Avenue NW and "G" Street NW, in the middle of the GWU campus. The sun is setting. The sidewalks are full of GWU students heading home from their last class. When the light turns green, I begin walking north on "G" Street.

As I walk down the street, I pass a worn-down, red brick, urban style school building. The school building and asphalt play-area are locked down behind an iron fence. Something makes me turn my head to the right.

Across the street is a one-story white building with a deli sign. Students are rushing in and out of the deli in a hurry. Just outside the deli, a homeless man is lying down on a bench. The students go out of their way to avoid him.

His hair and beard are both dirty gray. Both hair and beard are so long and so filthy that they are matted together, intertwined in a grotesque sweater of human hair extending down to his waist.

His khaki pants are stained yellow and brown. They give off a stench so foul, it makes me want to vomit. But I cannot keep from staring at him. He makes Aqualung look good. He is the grossest human being I've ever seen.

He sees me. This human port-a-potty is staring right at me. He waves at me urgently. He wants to speak to me. I don't want anything to do with him, but something makes me cross the street towards him. Crossing the street, I recoil from the gut-wrenching odor. I cannot go any further.

"There," the phlegm curdles in his throat as he speaks. "Go there."

A hand reddened and wrinkled beyond its years points to an old, two-story brownstone next door. The front door is an aluminum screen. It has a "T" on the front. Just like the first initial in my last name. The house number is 2163.

Then I woke up.

The dream is over. It is time to begin a new day. But as I get ready for my classes, the dream is as vivid in my mind as if I had just lived it in real life. Most dreams fade quickly once we wake. But not these "waking dreams." They can stay with you forever.

I follow my routine that day. Classes in the morning and afternoon. The lectures and the mental grind of concentrating on personal injury and contracts case law consume my attention. Somewhere along the way, my wonderful dream is forgotten amid a day of torts and contract law.

By the time my last class is over, I am mentally drained. I leave the law school in a mental fog. The next thing I know, I'm standing at a street corner waiting for the light to change to cross the street outside the law school.

I look up…

The street sign says 21st Avenue NW and "G" St. NW. Then déjà vu hits me in the face like brick. The dream! As soon as I remember having the dream, the strange presence that visited me in the Arizona desert is back again.

Just like that night in the Arizona desert, I have an invisible companion within my own mind. Too mentally drained by the day's studies to question its presence, I let it take over. It is one of the most important decisions of my adult life.

I walk down the street in a trance. But I already know what I will see.

Red brick school building behind a metal fence?

There it is…

Just as it was in my "waking dream." Half-way down the block, a red brick public school wrapped inside a ten-foot-high metal fence that whispers of vandalism and school children at risk from bad outside influences. Coal-black asphalt serves as the playground. What was it like attending classes there; or teaching there; playing there?

I keep walking. I am totally under the control of the presence in my mind. It is my body walking down the street, but I am no longer driving the "bus." I am just along for the ride. Caught in a fugue-state, I aware of what I am observing, yet somehow distanced from all of it. I am being led by the mind by something I know not what. I keep walking.

Deli in a small, one-story white building?

There it is…

GWU students scurry in and out of the tiny grocery. They are all in a hurry to get to the Metro station and their ride home.

By now, I know the script. It's all been preordained. I look for the homeless man.

My god! There he is ! It's really him! He is truly the ugliest human being I have ever seen.

My body walks on as though I'm just another student heading somewhere, like all the others. But, on the inside, I am totally freaking out. I even know what all the students call him. They don't know his real name, because none of them can stand his aroma long enough to ask him anything.

Just as in my dream, GWU students give him a very wide berth as they enter and leave the deli. But Aqualung doesn't move. He doesn't notice any of them or me. He's reading a book.

What book he would be reading? But the presence controlling my walking, wants nothing to do with him. I avoid him but cannot escape his smell. It's just as bad as it was in the dream.

Aqualung has a very peculiar odor. He smells of bourbon—a bourbon from hell. Its bouquet? A refined, blended stench of dried urine, sweat, feces and body odor; aged in very old khaki pants like a fine old, oaken barrel of aged vomit.

There it is!

2163 "G" Steet NW brownstone stares at me like the Cheshire cat in *Alice in Wonderland.* A mischievous grin grows on its aluminum storm-door in the form of a "T." This "T" was the last image I saw in my dream the night before. I was living a dream in real time.

Climbing up the twin sets of stairs to the walkup front-door was one of the most mind-blowing moments of my life. My knees shook so much I almost fell on my face.

At the door, I stopped to take a deep breath. My entire future life rests on what happens in the next sixty seconds, I tell myself. My hand raises to knock on the door, but my hand is shaking too much.

Deep breaths…try to relax.

A trembling hand knocks on the door. The door opens. Two young women greet me. I tell them my sad, pathetic tale. It only takes me three sentences. I can only say three sentences without screaming in desperation.

But one of them frowns at me the whole time. She makes no effort to hide her distaste. What could I have done to offend her?!

"Law students are assholes," she says sternly. "This is a graduate house. We're all family, here. Assholes aren't allowed."

My heart is crushed. My mind screams. After all the hope my dream created in my heart. I wanted to die. Just shoot me and get it over with.

"He doesn't look like typical law student to me," the other one counters. "He's just pathetic. We take pathetic. You'll have to live in the basement closet for two weeks until we get an open room…but if you can handle it—you're welcome."

I had stumbled upon the one and only group house GWU had for graduate students. This is the graduate house I live there the entire three years—except for the two weeks I spent at the Youth Hostel.

People of faith would call that a miracle. To be honest, it sure felt like it to me at the time. It was another moment in which my acceptance of a power external to mine, was proven to me.

Corny? You may think that. But when you've walked where I've walked, corny is ok. But for the presence that has visited me, I would have died more times than I have shared with you in this book. I am not embarrassed to tell you that the next three years of law school—and the many years that followed—have been years of giving thanks, where I tried my best to have earned the miracles I have been given.

There is a common greeting we often give to each other. It goes like this…

"How are you doing?" someone asks.

"Just living the dream," is the response.

The tone is facetious yet endearing. It is an effort to acknowledge how fragile our hopes and dreams exist in a very harsh, unrelenting world. But few, if any, probably ever really had a dream that literally changed their life overnight. Every time I am asked that question can honestly say I am living a real-life dream.

Unfortunately, since my graduation from law school, GWU tore my brownstone down to build a small park. I can say to my old alma mater that they made a huge mistake. The graduate house they tore down was the site of an intervention by the Universal Mind into the life of one poor, pitiful soul.

I cherish that day as much as the night I spent, as a child, in the Arizona desert. I can understand how people of faith would react to each of these experiences. For a long time, I felt the same way.

But our frames of reference are different. Time and thought have convinced me that religious faith had nothing to do with either experience. But I also know that something is "out there." There are only two alternatives. Either there is a race of beings so far advanced that they are no longer corporeal but pure energy, or we are dealing with the raw power of the Universal Mind.

Or it may be a combination of both. Perhaps a hidden race of non-physical beings possesses an understanding of the Universal Mind and have found a way to use it for their own purposes in a symbiotic relationship. Perhaps, the Universal Mind uses them as its messengers of good will and good deeds. I like that—Jedi Knights of the Universe. Maybe for some unknown reason they adopted us humans and have become our unofficial guardians. Maybe they just have a soft spot for the pathetic.

CHAPTER 16

Inspiration: As Evidence of Quantum Nonlocality

Let's go back to the dream I had about finding a place to live while in law school. What caused that dream? There can be only two logical possibilities. Either the dream originated within my own consciousness, or it did not.

Let's examine the first possibility. If the dream originated within my own mind, it had to have been the result of neurally-stored memories of life experience (this life or a past one whose memories have risen to conscious level) that were stimulated through either deductive or inductive reasoning. This reasoning then would have occurred to me as an idea from within my own mind.

There is the obvious flaw, however, to this possibility. I had never been down "G" Street NW before—at least not in this life. Therefore, there is no way I could have seen the graduate house located at address 2163 "G" St. Further, I had never contacted anyone from George Washington University about housing possibilities. The thought never occurred to me to do so.

Therefore, there is no possible way I could have ever, in this life, observed the "T" on the aluminum door, or the aluminum door, or the street number, or the deli, or the reeking homeless man, or the red brick public school across the street. Yet all of these images were part of my dream. How does that happen?

Both deductive and inductive reasoning require a base of knowledge from which to work. This knowledge, in the form of relevant experience, is the spark from which our mind synthesizes deductive and inductive reasoning into images we associate as ideas. Without past experience to serve as this spark, neither reasoning method is engaged within the mind. Therefore, no image. Therefore, no idea—and no dream.

But I had a dream. It was more than a dream. I had a specific vision of a future event—one that had not yet happened. I had a vision, as a waking dream, of me standing at the street corner, waiting for the light to change; walking down the street; walking past an urban-style public school; then turning to stare at a small deli—and the homeless man. Only then do I "see" the grad house with the aluminum storm-door with the "T" on it and a specific street address—2163.

To be clear, my brain obviously transmitted the images to my mind. I did dream the dream. My brain was engaged in the expression of the dream. However, my own neurons, my own neurally-stored memories, could not have been the source of the images. No physical connection existed between my own personal experience and the dream content.

We have no choice now but to accept as an objective fact that my brain lacked the capacity to create the images I "saw" during the dream—at least on its own. So, where did the dream images come from? Where did the vision originate? And who sent it to me?

To answer this question, we must dedicate ourselves to digging a little bit deeper into the realm of quantum mechanics in general and specifically, that expression of quantum reality known as 'nonlocality."

Do not worry. You will not need to know about "Hilbert spaces," brakets, linear operators or any other complicated aspects of quantum mechanics. Just keep in mind our earlier discussion on entanglement and nonlocality. Simply put, two or more photons "joined" by an electromagnetic field are called "entangled." Entangled photons demonstrate the amazing ability to communicate with each other instantly, no matter how far apart they become.

Can entangled photons communicate faster than the speed of light?

We know now that the answer is yes. Light speed does not come close to instantaneous communication which entangled photons demonstrate. But all this means is that entangled photons share a means of communication which is must faster than they can travel individually.

This should not be difficult to accept. Sound waves travel faster than we can walk or run. Instead of running over to you to tell you something, I can yell to you what I need to say much quicker. All this shows is that the process by which photons travel is classical mechanical physics while their communication system is quantum mechanical physics.

Why?

The reason is quantum nonlocality. The original 1935 paper by Einstein, Podolsky and Rosen (EPR) theorizing that entangled photons shared "spooky" action at a distance caused much debate and research in the following decades. Then in 1964, John Stewart Bell, a physicist working for CERN (European Organization for Nuclear Research), published the first of a set of papers which mathematically proved Einstein and his team were right.

By the way, remember that name CERN. It will pop up later. And it will blow your mind—as will its connection to physicist John Stewart Bell.

"Bell's Theorem," as Bell's mathematical proof has been called, asserts that if "certain predictions of quantum theory are correct, then our world is non-local. 'Non-local' means that there exist interactions between events that are too far apart in space and too close together in time for the events to be connected even by signals moving at the speed of light."

Source: S. Goldstein, T. Norsen, et al, "Bell's Theorem," Scholarpedia, 2011.

Bell used mathematical formulae to show Einstein's theory that quantum mechanics allowed for electromagnetic energy to exhibit what he and his colleagues called "spooky" behavior had a scientific backbone to it. Repeated, independent scientific experiments since Bell's papers have proven both he and the Einstein team were right.

"The outcomes of the Bell tests provide dramatic confirmations of the *prima facie* entanglement of many quantum states of systems consisting of 2 or more constituents."

Source: Myrvold, Wayne, Marco Genovese, and Abner Shimony, "Bell's Theorem," The Stanford Encyclopedia of Philosophy, (Fall 2021 Edition), E.N. Zalta, (ed.)

The universe is full of information. That's what it was all about in the beginning. It is the same now and will be forever and ever.

A universal ocean of quantum-entangled information exists inside each of us. How can it not be? Our existence as beings comes directly from this initial flash. We are the children of the Big Bang. We are the children of this photonic "ocean" of intelligence. This "ocean" flows around us, even through us—every moment of every day, every year, of every life we ever have lived.

Why have we no proof Reincarnation exists? That is the question most people want to ask. The answer is simple; to the contrary, there is proof Reincarnation exists—albeit indirect.

To answer this, we must analyze how the human brain has evolved. Evolution has developed the brain to perform very specific tasks. The brain receives sensory data; analyzes that data to determine its deterministic relevance to us on a series of "yes-no" spectra: threat-nothreat/food-nofood/pleasant-unpleasant, etc. It then moderates our response to that stimulus, encodes the event with an appropriately weighted emotional label and then files the experience away for future use as memory.

In this way, the brain operates in tune with what physicists term Newtonian physics or classical mechanics. Isaac Newton's laws of motion, momentum and inertia are based solely upon what we humans perceive with our limited physical senses. Evolution developed our brain to function in such a way as to interact with our physical environment using our physical-universe-adapted senses to survive.

Our human relationship to the perception of classical mechanics is built into human and animal DNA as instinctive, reflex reactions to the environment. Why is it that when a predator, searching for food, will chase its prey, only to stop immediately when the prey's speed, and its distance from the predator, makes the prey uncatchable? This is the predator's evolutionary instincts kicking in, telling it that the chase is a waste of time and energy that would be better spent on a slower animal.

But the universe of quantum mechanics is totally alien to our senses. The quantum world is comprised of the fundamental building blocks of the universe—objects so small, in both size and mass, that humans are incapable of interacting with them on a sensory. As the acclaimed physicist, author, lecturer, and YouTuber, Dr. Leonard Susskind, Ph.D. stated in his best-selling book, *Quantum Mechanics: The Theoretical Minimum*:

"Quantum abstractions are fundamentally different from classical ones. For example, we'll see that the idea of a state in quantum mechanics is conceptually very different from its classical counterpart. States are represented by different mathematical objects and have a different logical structure.

In the classical world, the relationship between the state of a system and the result of a measurement on that system is very straight-forward. In fact, it's trivial. The labels that describe a state (the position and momentum of a particle, for example) are the same labels that characterize measurements of that state. To put it another way, one can perform an experiment to determine the state of a system. In the quantum world, this is not true. States and measurements are two different things, and the relationship between them is subtle and nonintuitive."

Source: Quantum Mechanics: The Theoretical Minimum, Dr. Leonard Susskind & Art Friedman, Basic Books, New York, 2014.

This relationship between our brain and the quantum universe of fundamental particles and energies, such as the photon, is nonintuitive because the evolutionary development of our human brain did not account for any need to interact with the quantum universe. Our only interaction with the quantum universe is indirect and, I must say, basic.

We use the abstract world of mathematics to hypothesize how quantum mechanics operates. Once scientists develop what they believe to be a mathematically correct theory of how a fundamental particle might function, they use giant particle accelerators, powered by huge magnets, to cause these particles of fundamental matter together to see whether the theory is accurate or not.

Pretty basic, wouldn't you say?

How can humans, limited by evolution to a classical mechanical understanding of the universe, be capable of such insights into the quantum mechanical universe as to be able to theorize its existence and processes?

Is it merely a matter of inductive or deductive reasoning? No, it is not. Both inductive and deductive reasoning is based on our relationship with classical mechanics and experience—what the brain would consider as memory. We humans have NO prior experiential knowledge of quantum mechanical effects.

We cannot see sub-atomic particles with our naked eye. We cannot see photons as waves or individual particles. We only see visible light. Prior to Ben Franklin and his kite, our only knowledge of electricity was that it came from the god Thor.

Classical mechanics is intuitive. Classical mechanics is about what we can feel, touch, hear, smell and taste. We "live" classical mechanics and apply its laws every day of our lives. We see it, feel it, touch it, hear it, smell it, and taste it. Rooted deep in our DNA, classical mechanics uses physical processes we humans and our evolutionary ancestors have employed to survive for millions, if not billions of years.

Quantum mechanics is not intuitive. Neither we, nor our evolutionary ancestors have knowingly used its properties. We have no conscious connection to what we cannot see, touch, smell, hear, or taste.

Can we see, touch, smell, hear or taste an electron, or a single photon? No, we cannot. Then how could physicists, whose only academic experience was with classical mechanical laws, suddenly begin to hypothesize quantum mechanical existence in the first place?

That is the big question, isn't it? How do albeit great minds conceive of what is hidden when they have no prior conscious exposure to its existence? Remember, to do something, one must think of it first.

We use either inductive or deductive reasoning, aided by observation of physical phenomena in our environment (or through scientific experiments). Deductive reasoning is simple. You trip and fall running down a flight of stairs and you can easily deduce from your experience that running down a flight of stairs involves risk of harm. Two plus two equals four. That is deduction. Simple.

> "Deductive reasoning is a basic form of valid reasoning. Deductive reasoning, or deduction, starts out with a general statement, or hypothesis, and examines the possibilities to reach a specific, logical conclusion…The scientific method uses deduction to test hypotheses and theories…"
>
> "…In inductive inference, we go from the specific to the general. We make many observations, discern a pattern, make a generalization, and infer an explanation or a theory…In science, there is a constant interplay between inductive inference (based on observations) and deductive inference (based on theory) until we get closer and closer to the 'truth.'"
>
> Source: A. Bradford, "Deductive Reasoning vs. Inductive Reasoning," LiveScience, at livescience.com, 2017.

Both deductive and inductive reasoning are classical mechanical operations of the mind as it works through problems. But this type of reasoning does not work with quantum mechanical processes that we have no sensory experience with.

So how could Einstein and other physicists, who have made discoveries in quantum mechanical processes, come up with theoretical visions of a quantum reality despite lacking any frame of reference with which to conceive of the system they are attempting to describe?

Again, the answer must be the human mind's ability to connect nonlocally with a universal consciousness.

Professor Susskind is one of the greatest physicists of all time and one of the greatest minds of the last 100 years. Very few humans have been able to prove famed physicist Stephen Hawking wrong about anything. But Professor Susskind proved Hawking's theory that black holes destroyed information was completely wrong. In an intellectual *tour de force,* Professor Susskind proved information is not destroyed, only transformed from one state to another—just as the Law of Conservation of Information predicts.

Professor Susskind's views on how quantum mechanics are nonintuitive bear reading:

"Classical mechanics is intuitive; things move in predictable ways. An experienced ballplayer can take a quick look at a fly ball, and from its location and its velocity, know where to run in order to be there just in time to catch the ball...There is an obvious reason why classical mechanics is intuitive; humans, and animals before them, have been using it many times every day for survival. But no one has ever used quantum mechanics before the twentieth century.

Quantum mechanics describes things so small that they are completely beyond the range of human senses. So, it stands to reason that we did not evolve an intuition for the quantum world. The only way we can comprehend it is by rewiring our intuitions with abstract mathematics. Fortunately, for some odd reason, we did evolve the capacity for such rewiring."

Source: L. Susskind and A. Friedman, Quantum Mechanics: The Theoretical Minimum, Basic Books, New York, 2014.

That is Professor Susskind's attempt to explain the unexplainable. How do humans conceptualize what they have never experienced? His answer is "abstract mathematics."

That is the same as saying we humans use deductive and inductive reasoning to understand the results of scientific experiments. But this does not explain how scientists create the abstract mathematics or the experiments they use to test their theories.

Abstract mathematics is just a mathematical language scientists create to explain and predict behavior of variables within any system they are studying. Mathematics has a logic and internal consistency scientists have found invaluable in understanding the universe.

Eventually, scientists will develop the mathematical formulae and equations to explain Reincarnation and the Universal Mind. This is inevitable.

But humans did not invent mathematics. We learned it. From whom or what did we learn of it? This question begs for an answer. An answer that cannot be found in any church.

Principles of mathematics and the relationships between matter and energy were determined just after the big bang. All scientists have done is discover its principles and learn, over the centuries, to apply them to different physical phenomena.

When we get down to it, the hardest part of science is not the deductive or even the inductive reasoning that occurs when one theorizes. The most difficult part of science is conceiving the experiment that tests the validity of a theory. An experiment is a test of a theory. But it takes a creative leap of imagination and inspiration to make this happen.

It is difficult enough to design experiments to test a theory one does not fully understand in classical mechanics. Imagine how difficult it would be to design an experiment in quantum mechanics where the researcher has no intuitive grasp of the subject.

What do artists, writers, actors, and craftsmen have in common with scientists? They all create. Creativity is more than deductive and inductive reasoning. Each depends upon a "leap of logic" called inspiration.

Whether it was Eratosthenes measuring the world in the 3rd century, B.C., William Harvey discovering blood circulation in 1628, Sir Isaac Newton discovering gravity, Gregor Mendyl discovering the fundamental values of genetic heredity in 1855, Marie Curie defining radioactivity, Thomas Young's discovery that photons were not only individual particles but also moved in waves, or Einstein's $E = mc^2$; all these great minds needed inspiration to both design the experiments which led to their discoveries and to understand the results.

Thomas Young developed an experiment which allowed sunlight to shine onto a mirror and then into a small hole. He then "split" this light beam in half using an ordinary piece of paper and observed how the light cast what he called an "interference pattern" of colored and dark lines on a light-reactive screen. He reasoned that the only way this pattern could have occurred is if light moved as a wave.

How did Young think to split the light beam? How did he suspect that by splitting the light beam, anything would happen? This required as much inspiration as it took to come up with the idea of light duality (particle-wave) itself. It certainly did not come from any prior experiment Young performed. He had never experimented before with particle-wave duality. So it is a complete mystery.

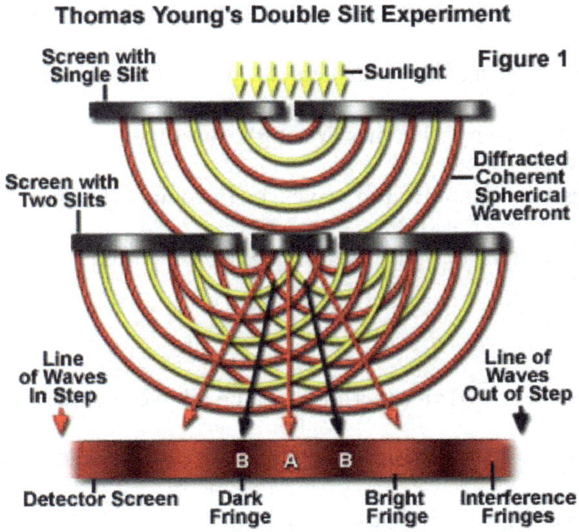

Illustration No. 26. Diagram of Young's double-slit experiment. copyright 2022

In Illustration No. 26 above, when Dr. Young examined the detector plates he found an alternating pattern of light and dark areas. The only explanation for this phenomenon is that the double-slit creates two competing waves that at certain points cancel each other out. This thought process is deduction.

But deduction cannot explain his choice of such a "double-slit" technique to use in the experiment. That took a combination of inductive reasoning, based upon experience and training, and inspiration.

Both deductive and inductive reasoning rely upon human experience. But human experience does not explain the inspiration that occurs during art, music, creative writing or scientific discovery. That leaves the source of such insight to be an unknown source of information located outside the brain.

What could this be?

Is it "inspiration?"

Yes and no.

The answer is "yes" because in a relative universe, what we find to be new to our personal experiential existence is by definition, "inspired."

We each interact with information separately and uniquely to all others. That is the foundation of individuality with which Reincarnation was created by evolution to conserve—along with the unique information we hold within our memories which have formed that individuality.

The answer is also "no."

We consider unexpected information to be "inspiration." Such information appears to us to be "inspired" because it does not appear to us as the result of deductive or inductive reasoning based upon our past experience and training. However, this does not mean others have not been aware of the same information previously.

History is full of examples of scientists and philosophers in different parts of the world each coming up with theories and ideas at around the same time without knowing someone else was also thinking the same or similar thoughts. This phenomenon is not rare. Indeed, it has been recognized as a common occurrence in both scientific discoveries as well as inventions.

"Multiple independent discovery" or "independent invention," as this phenomenon is named, does not mean that separate research papers and articles are identical. What it does mean is that generally identical discoveries arise in the minds of separate scientific researchers, inventors, and entire cultures even though totally isolated from each other.

Robert K. Merton, Columbia University professor considered to be the founding father of modern sociology, defined such "multiples" to occur when similar or even identical discoveries are made by scientists and inventors who whose "origin" theory and subsequent research is independent from all others.

"Sometimes the discoveries are simultaneous or almost so; sometimes a scientist will make a new discovery which, unknown to him, somebody else has made years before."

Source: R.K. Merton, "The Sociology of Science: Theoretical and Empirical Investigations," The University of Chicago Press, Chicago, 1973.

Accepted examples of this "multiple independent discovery" phenomenon include:

1. Calculus by Isaac Newton, Gottfried Wilhelm Leibniz, among others.
2. Oxygen element by Scheele, Priestley, Lavoisier and others.
3. Theory of evolution by Darwin and Wallace.
4. The blast furnace by multiple inventors around the world (China, Europe and Africa).
5. The crossbow invented independently in China, Greece, Africa and North America.
6. Magnetism discovered independently in Greece, China and India.

Source: "Multiple Discovery," Wikipedia, (no date).

According to Professor Merton, "multiple independent discovery" is not limited to only a few historic instances involving acknowledged "giants" in their respective fields but is a common pattern in scientific discovery. Furthermore, this phenomenon is as common in the arts and literature as it is in science.

More than one novelist, playwright, or screenwriter can come up with the same plot and characters, independently from others, yet write different versions of each. Multiple artists and songwriters can be inspired by the same general idea, yet use their individual experience to express their vision in different ways.

What do these examples of "multiple independent discovery" teach us? It makes no difference whether the application of the shared inspiration is reflected in science, engineering, or the arts. The core source in each remains the same. Each is an expression of inspiration or the "creative process."

Where does this commonality of inspiration originate?

As we must now accept the reality that the universe is interconnected through non-locality with everything that exists within it (including all life), we must now contemplate the consequences of such a realization—expanded consciousness connected at a fundamental level to a universal mind.

Inspiration and the "creative process" is not limited to the intellectual or the academic. All of us have experienced the phenomenon of shared inspiration in our daily lives. Communion with the Universal Mind is an ability we all possess because we are all part of the same universe.

In his 2015 Explore article, "The Implications of Nonlocality," Stephan A. Schwartz writes:

"The corollary to the idea of nonlocal consciousness is that all life is interconnected and interdependent, that we are part of a matrix of life, but even more fundamentally, that space time itself arises from consciousness, not consciousness from space time."

Source: S. A. Schwartz, "The Implications of Nonlocality," Explore, January-February, p.8, (2015).

If Schwartz is correct and space-time arises out of consciousness, then from where does consciousness arise?

To answer that question, we must take Schwartz's vision and go beyond it. We have established that my Washington, D.C. dreams had no connection with anything I personally experienced in this life. Therefore, the visions and images I received in my Washington, D.C. dream clearly came from "outside" my mind.

Some will attempt to undermine the dream I lived as merely anecdotal evidence of nonlocality in human consciousness. But I was there. I experienced it.

I am convinced my dream was a pure "non-localized" communication based upon quantum entanglement principles. My intense "prayer" (thought wave) created a quantum worm-hole in space-time allowing the Universal Mind to directly communicate with my subconscious dream-state in real time. But that is not all the Universal Mind did.

The Universal Mind not only gave me the dream to inform me what was about to happen. It MADE it happen. The Universal Mind acted. Therefore the Universal Mind must be capable of acting independently of our will or control.

Scary, huh?

CHAPTER 17

Swimming the Photon Sea

I am just a cup of water until poured into the ocean. Then, there is no more me, just the endless sea. My five senses tell me that I am separate from everyone and everything else. All I can remember is being separate from everything, even the universe itself.

But our senses create an illusion. The illusion is that we are separate from each other; separate from everything. But the reality is much, much different. We are all swimming in an endless sea of consciousness and light.

Einstein's theory of relativity appears to confirm this. We each assume we have separate individuality because we each experience a separate perspective of existence and we each possess memories of an existence separate from all else. This separate perspective of existence creates an individual frame of reference that is separate and unique from everyone else's.

Individuality is not the only perspective of existential awareness. Group-mind was also an evolutionary option. Individuality or the memory of separate existence is likely a function of evolution's effort to diversify experience to maximize survival of species.

Considering that the Universal Mind is in essence a "group-mind" of universally entangled photons, it would seem logical that the Universal Mind would use evolution to mimic this function as well. That the Universal Mind did not make this choice presents an interesting issue.

What purpose would the Universal Mind have in creating individuality as an evolutionary adaptation of species development? The consequences of such a choice were dramatically destructive. How many people commit suicide every year because they feel "lost" and "alone?" Would they have felt so alone had they been aware they share thoughts with the Universal Mind and through that Mind, everyone else?

Remember my journey into the Arizona desert and the "voice" that spoke to me in my mind? Remember my dream when I was in law school? These events cannot be explained by classical mechanics. Neither can they be explained away as the irrational ramblings of a crazy old man. A brief look at my bio at the end of this book should dispel any notion of that.

What this means is that while on a grossly physical and material perspective we are "separate" from other forms of matter, this separateness is an illusion. In reality, our collective consciousnesses are interconnected through quantum entangled photons which are themselves quantumly entangled. This quantum entanglement of photonic information is the Universal Mind. It connects all life and all consciousness in the universe.

Earlier in this book, we discussed what is called the Cosmic Microwave Background (CMB). We even used an illustration to show how this radiation looked immediately after the "decoupling era" ended (Illustration No. 1).

The CMB was discovered by two Bell Lab scientists, Robert Wilson and Arno Penzias in 1964. Bell Lab (the research entity of the old Bell Telephone Company) were puzzled what was creating the strange buzzing sound picked up by its Holmdale Horn antenna in New Jersey.

Wilson and Penzias were given the task of trying to eliminate the background static. They tried pointing the antenna in different directions. But no matter where the antenna was pointed, the buzzing continued. The buzzing seemed to be coming from everywhere at the same time. What could do that?

What Wilson and Penzias had discovered was the "echo" left by the electromagnetic energy released from the Big Bang. Not only was this proof that the universe began from a single event (the Big Bang), but it was also evidence of something just as significant.

Source: M. Wall, "Cosmic Anniversary: Big Bang Echo Discovered 50 Years Ago Today," Space at space.com, 2014.

This CMB "echo" appeared to come from all directions with equal intensity. What does this mean? It means that electromagnetic energy from the Big Bang is everywhere in universe with equal density. There is no variation in the signal, no matter where we look.

What is the cause?

We must begin by recreating the universe after the "decoupling era." Individual photons are released from the plasma field. This photon release has two effects.

1. The universe becomes filled with an ocean of individual photon particles. These photons are traveling ahead of us and exist between us and the edge of the universe as well as exist between us and the universe's origin.

2. Photons also are absorbed by the universe's expanding lithium skin.

What exists is essentially a universe filled with photon particles in an expanding hydrogen-helium gas. Photons have two properties. They act as individual particles or "quanta" of electromagnetic energy. But photons also act together to create energy waves.

Is there a way we can visualize how this photonic environment interacts with us in the universe? While our evolution does not give us an intuitive "feel" for the photonic ocean that fills the universe, can we conceive of a way to compensate and create a visualization?

Imagine it's your sixth birthday. Your parents take you to a children's amusement park. In the park is a bounce house with thousands of rubber balls. Each is separate but touching others adjacent to them as shown below:

Illustration No. 27. The Bounce House. Source: Creative Commons copyright 2022

The bounce house is full of these rubber balls. They are so tightly packed together; you don't think you can even step inside. You don't think you can fit inside but you try anyway. But when you step inside, your eyes widen in amazement.

You step inside the bounce house, you can see the rubber balls, but you cannot feel them. Your body passes right on through them. You only feel a warm glow, like touching a light bulb that is turned on. Each ball glows with the changing colors of the electromagnetic spectrum.

As you enter the bounce house, the balls you touch absorb energy from your body's momentum, heat, and electromagnetic field. As their energy levels change, so too do their vibration levels and their colors.

The energy from your body passes on to the balls you touch or come closest to touching. These turn blindingly white. The balls they touch, as a result of the transfer of energy from you through the bounce house, change color to bright yellow to indicate their change of energy level—not as energetic as the balls you actually touch, but almost.

As you move through the bounce house, the balls vibrate and change color according to how close they are to you. The balls closest to you are pure white and they vibrate much faster than any of the other balls. But as the distance from you to any ball increases, the balls both change color and vibration level.

The balls change from white to yellow, then to orange, red, green, blue, and then to darker shades of violet. Their vibration levels also change according to the relative loss of electromagnetic energy as the distance between you and each ball increases.

The balls furthest from you are black and do not move at all. They have no energy. But as you come near, some of your electromagnetic energy finally reaches them. Then, they too, begin to vibrate faster and faster and change color (violet to blue, then blue to green, then green to red) as they absorb more and more energy.

Rippling waves of color flow through the bounce house with you at its center. It is as beautiful as it is dynamic.

Now, replace the bounce house with the infinite, physical emptiness of the universe and replace the stationary massless rubber balls with massless photons coming at you from all directions—each one traveling at 300,000 kms/sec.

Don't worry. They can't hurt you. Being massless, the photons pass right through your skin, into your body. Most pass right on through you, without you ever feeling any of them. But some are absorbed into your body, even into your brain.

The information they carry is transferred into your neurons and synapses. Imagine all that information incorporated into your being, into your very consciousness. You remain blissfully unaware of all of it, your entire life.

Visualize this and you come close to visualizing how we exist with an ocean of photons surrounding us and flowing through us every second all our lives. This is the photon ocean that has filled the universe for 13 billion years.

Now let's focus on how photons transfer energy or information. Photons exist as separate, discrete units that absorb, carry and transfer information in succession. This can be illustrated in the following way: fill a box with small, soft balls than can readily absorb and transfer energy. Small gummies, erasers, rubber balls or something similar work best.

Each ball touches several other balls, which in turn touch several others, and so on throughout the box. Now, use your finger and flick one of the balls. What happens?

The first ball you touch transfers the energy from your finger-flick to the balls that it touches. These balls then strike the balls that they touch, and so on. This transference expands out and away from the initial ball in what appears to be an expanding wave. Each ball symbolizes a single photon. They exist individually as electromagnetic particles, while simultaneously moving together as wave energy.

This duality is one of the weirder aspects of quantum mechanical processes. On the other hand, we see similar processes in our daily lives and take them for granted.

The river, streams, lakes, and oceans of our world are seen from above as one mass of undivided water. But turn on your faucet and you can narrow the focus of the mass down into a discrete drop. Is that single drop any less "water" than the whole from which it came? Are any of its significant properties any less viable because it has been reduced to its "theoretical minimum?"

The answer to each is, of course, no.

CHAPTER 18

Meet Your Maker

Let's go back in time…

Who or what was communicating with me that night in the Arizona desert? Who or what gave me the vision of floating in space beside Jupiter as it shared with me its compassion for a suffering humanity, lost and alone in the darkness? Who or what gave me the "waking dream" in Washington, D.C. that saved my law career?

Was I contacted by advanced beings through telepathy and mind power? Or was I communicating directly with the Universal Mind?

Whether we are interacting with a race of beings far more advanced than we, or we are interacting with a Universal Mind, the initial questions are the same. What are their or its intentions and how do we communicate with them—or it?

This assumes, of course, that in either case, there is a willingness on the other end for mutual communication with us. It is difficult to imagine what a species, able to control telepathy, might have enough in common with pathetic humans to want to communicate with us.

Do you have a pet dog? How often do you play with him/her? How often do you communicate with him/her? Do you discuss politics with your pet? How often do you debate with your pet about which is best—idealism or pragmatism?

See my point? We share no frame of reference with these alien beings. We have nothing that they believe has any value to them. Most likely, their only interest in us is based upon only two motives. Their interest in us is either based solely on self-interest or a combination of self-interest and ethnocentric-altruism.

What is ethnocentric-altruism?

An example of ethnocentrism:

This advanced race of beings obviously believes in its fundamental superiority to us. They have all the answers. We have none. They would never consider sacrificing themselves or their race to save our own. This is ethnocentrism.

An example of ethnocentric-altruism:

If their interactions are at all motivated by any compassion or sense of altruism, how they express this compassion towards us would be limited by what their culture deems acceptable given their subjective frame of reference. This is ethnocentric-altruism or "self-centered compassion."

The closest any human relationship comes to this "self-centered compassion" is not parenting or even pet ownership. Parents are willing to sacrifice themselves for their children. So are many pet owners willing to risk their lives for their pets.

No, if there is an alien race interested in us at all, the closest any human relationship comes to what this alien race thinks of us is foreign aid. We send foreign aid to those nations we wish to be our allies. But that does not mean we will risk our own safety just to save them from threats. This is ethnocentric-altruism or "self-centered compassion."

Whatever one's species, the only way to survive and prosper is to effectively process the information your environment gives you as sensory stimuli and make sound and productive choices in return. The ability to effectively process information is now the most important skill-set anyone can have. Without the intelligence that effectively processes information, a person is like a ship searching for a safe harbor of truth without a rudder or engine; adrift in the turbulent, stormy seas of lies, misinformation, propaganda and hidden-agendas.

For better or worse, we humans define ourselves by what we do. Let's try an experiment. Put yourself in the position of each of the following occupational types and answer the following questions as best you can.

	A	B
1.	Surgeon	Fast-food worker.
2.	Trial lawyer	Long-haul truck driver.
3.	Quantum physicist	Unskilled construction laborer

What do the occupations in Group **A** have in common with those in Group **B**? How are they different? Lastly, which occupations are most likely to exist fifty years from now and why?

I think you should be able to answer those questions without help from me. But I will give you this much help; the answer is a logical argument. This argument goes like this:

(1) future human development is occupationally dependent;
(2) occupational development is educationally dependent;
(3) educational development is dependent upon information-processing; and
(4) the ability to process information is genetically/environmentally-dependent.

For anyone in Group **B** to carry on a meaningful conversation with anyone in Group **A**, he or she must climb an IQ-rope such as we have just described above. Imagine the struggle an unskilled worker has in order to "climb" this IQ-rope to carry on a meaningful conversation with anyone from Group **A**.

How likely is it that anyone in Group "B" will ever interact socially with anyone in Group **A**? The only time a fast-food worker interacts with a surgeon is when the fast-food worker needs an operation.

When does a long-haul truck driver interact with a trial lawyer? When the trucker is arrested for DUI or domestic battery and wants a plea deal or a trial. Who can relate with a quantum physicist?

For centuries, physical labor was the key to success and power. How well can one fight? How well can one hunt? How well can one farm, fish, build or repair equipment? For thousands of years these were the paths to power and a privileged life—unless one was a celebrated artist, musician or scholar.

But all that has changed forever. The industrial revolution and scientific-technological advances since the late 1800s dramatically changed the skills needed for humans to succeed. One can say that the industrial revolution was a mutation within the human genome. It has forever altered human development.

Information-processing is now the future of human development. Look around you at the people you interact with. Do they have a college degree? Do they have an occupation that is safe from being replaced by robots or even artificial intelligence? If they do not, within the next fifty years, their jobs will all disappear—along with the income and life-style those jobs provided.

Worse than this, however, is the effect such a lack of education of the parents will have upon their children and succeeding generations. During my twenty-five years as a criminal defense trial lawyer with the Office of Public Defender, I specialized in defending those accused who were too poor to afford their own lawyers.

Over my twenty-five years as an Assistant Public Defender, I represented thousands of clients. They were all indigent—too poor to hire their own attorney. What caused them to be poor, especially in a nation where opportunity to become a success is so freely available (unless, of course, one is trapped in the vise of institutional racism).

Because I suspected lack of education was one of the main causes crime among the poor, I kept an informal tally of my Public Defender clients' educational backgrounds. Many of the "crimes" my impoverished clients were accused of were the result of their being unable to interact with governmental authority (license registration, traffic offenses, hunting and fishing violations, gun registration, local ordinance violations, etc.).

Many of these clients lacked the ability to function successfully within our modern, high-tech, and information-loaded society.

1. 95% had not graduated from college.
2. 90% had not attended college.
3. 80% had not graduated high school.
4. 40% had not completed 10th grade.
5. 20% had not started 9th grade.

(Approximated percentages from personal experience).

Successful achievement in school is more than being able to memorize and recite. Pets can do that. Your pet dog can remember a verbal or physical cue and do what you have trained him or her to do. Success in school requires the evolution-dependent ability to process information, using inductive and deductive reasoning to comprehend what is being memorized. That is what "learning" is all about.

Sadly, a considerable proportion of humans apparently lack even that basic level of evolution-dependent intelligence. It is not their fault. They are victims of genetic history that failed to adapt to our modern, information-dependent world.

Less than a hundred years ago, their genetic history would have enabled them to succeed as farmers, ranchers, hunter-gatherers, fishermen, and craftsmen. It used to be that such skills would make one—if not rich, at least able to raise a family with better opportunities for growth than they had when they started out.

But that world is forever gone. Steven Jobs, Bill Gates, and others like them have seen to that. The soil, land, sea, and forge no longer provide wealth and power for individuals. Their life-styles have become reduced to subsistence income only. And their future is even more grim.

Now, let's take these information-processing skills forward 10,000 years; 100,000 years; a million years. Then, look back in time to where we exist today. Do you see how difficult it will be for us to communicate with whatever these advanced beings are—if they exist around us now? They are too far advanced to have any desire for meaningful communication with us.

Beings that much more advanced than us, who are able to exist as noncorporeal entities would have no reason to contact us. We would be as relevant to their existence as a delivery truck driver at conference of surgeons.

We do not know how long it takes for intelligent life to develop on any planet other than our own. We have no basis for comparison—yet. We are the only species we know that exists.

We do not know for certain what exists alongside of us in the nonphysical universe. Dark matter…dark energy…it's all unknown to us. But one thing is certain from what we know.

There is a universe. There is also a Universal Mind and it is waiting for us.

I have shared with you four events from my life that are clearly out of the ordinary: (1) my near-death-experience in the hospital; (2) the telepathic "voice" in the desert; (3) my lunch-time vision being in space looking back at humanity on Earth; and my GWU law school dream. Each of these, and countless other experiences the Universal Mind has shared with me over the 70-plus years of my life have given me a unique insight into what the Universal Mind is.

To understand how the Universal Mind interacts with us we must understand its nature. What is it made of? What is its frame of reference? Until we can answer these questions, we have no business hypothesizing what or why it does what it does. Let's speculate intelligently about what its nature must be…

1. The Universal Mind must be an organic-consciousness. It measures itself, its surroundings, and its environment. It is infinitely intelligent, with infinite memory storage and awareness of all its parts at some level. Therefore, we must consider it to be a life-form—the most unique life-form in all the universe.

2, It has its own agenda. It is capable of acting independently of our will.

3. As pure, nonphysical consciousness, it cannot act on its own. However, through photonic entanglement and nonlocality, it knows all, sees all, and communicates telepathically with any living being anywhere in the universe instantly through quantum entanglement and nonlocality—when it chooses. It can certainly affect the lives of all species and individuals—when it chooses. It certainly did so with me.

4. The Universal Mind can "feel" our hopes and aspirations, as well as our pain and suffering. It is capable of doing this on both the species level and also on the individual level—when it chooses. "Miracles" happen. Miracles are spontaneous expressions of quantum nonlocality.

4. Then why doesn't the Universal Mind save us from ourselves? Only power-hungry, ego-maniacal men create a god that plays savior and a religion to force obedience to its false-powers. Such men need such a power-base for their intended empires, large and small. The Universal Mind does not save us from ourselves because it cannot. It must obey the Natural Laws and relationships between matter and energy that were determined and fixed during the "decoupling" era of the early universe—just like we must. Without those laws and relationships being determined and fixed, there would be no Universal Mind—or us. The Universal Mind is as trapped within its own Rules of Creation (including evolution and natural selection) as we are. The program is running. The Natural Laws and relationships are algorithms preventing any changes, hacking or other alterations to its core. Einstein once said, "God does not play dice." Einstein failed to explain why not. The "why" is obvious…the "dice" have been deleted. The program now runs itself.

5. But the Universal Mind is capable of intervention—when it chooses. The Universal Mind communicates through the universal-consciousness its shares with all intelligent species through nonlocality. Our dreams, our inspirational moments, our thoughts, our meditations, our prayers are ALL heard at some level. And, if it chooses, it is able to affect our behavior by altering our thoughts to change the course of our lives.

6. The Universal Mind suffers from two specific mental illnesses. It is both neurotic and a control-freak. It suffers from an obsessive-compulsion to control ALL. There are consequences:

(1) It is self-contradictory. It is objective, but not. It is subjective, but not. It is both all and nothing—at the same time. As it can perceive subjective stimuli it knows suffering of its many different species but cannot "save" anyone directly.

(2) There are consequences to the above. By definition, the Universal Mind is always in conflict with itself over what it wants versus what it perceives at any one time. Did the Universal Mind "save" the dinosaurs from their extinction? Do you really believe the Universal Mind is "attached" to our own species any more or less than the dinosaurs?

We may have more intelligence than they, but that distinction alone will not save us should the Universal Mind decide the human race has been run—and lost.

7. The Universal Mind is not a god. It does not require your worship, or your money. You need not kneel before it. It is a life-force that exists through its physical manifestations—life. It exists through us. Meditation is not the same as worship.

Is it possible for a Universal Mind to "turn" against one of its own creations and allow it to be destroyed by simply refusing to intervene? Could the Universal Mind even act with premeditation to destroy entire races of species? Of course, it can. It's already set the path to our inevitable destruction in motion.

We find it difficult, if not impossible, to think of time and space as interrelated dimensions of existence. It is too abstract a concept. But we accept intellectually that they must be. The universe is both time and space simultaneously.

How can the Universal Mind consider itself as anything other than as one organic being? Concepts rooted in division of perspectives like "us and them" or "objective versus subjective" or "we and they" or "I and you" must be completely alien to its consciousness.

The Universal Mind can no more think of you or I as individuals, separate from itself than you or I can consider our arms as separate entities. Does the Universal Mind respond when a massive solar flare release billions of gamma rays upon a solar system and doom all life living within it?

No.

In 4.5 billion years, give or take a few million, our Milky Way galaxy will collide with the Andromeda galaxy. The result will be catastrophic for both. Billions of star systems will be destroyed. Megabillions of planets will be obliterated. All life on two colossal galaxies will cease to exist. How will the Universal Mind respond to that?

Will the Universal Mind "feel" compassion at such an immense loss? Will it miss us when our civilization becomes extinct?

No.

The Universal Mind will instead observe the catastrophe with the thought:

"Behold, the universe is in balance and all is progressing as intended."

The Universal Mind has had billions of years to prevent or change the trajectories of both galaxies to save two entire galactic civilizations from total annihilation. Nothing has changed. Our two galaxies are still on a path toward mutual destruction. But we know such intervention was made impossible by the Natural Laws it allowed to be set in place.

Either way you slice it up, the answer will be the same. The tragic trajectories of the Milky Way and Andromeda galaxies were set upon their mutual paths to mutual destruction at the dawn of the universe. The coming collision is no accident. Perhaps the knowledge that something new will result from the collision will yield us some consolation. But the universe needs none.

The Universal Mind may "hear" our brief and futile cries of agony at our impending doom, but will it do anything to stop it? When the next killer-asteroid approaches Earth, will the Universal Mind do for us what it refused to do for the dinosaurs? Will our prayers for a dream of survival be answered?

Do we feel pain when we scratch our arms? Of course, we do. Do we put a bandage over it? Yes, but the skin must heal on its own. If we, as a civilization, cannot learn to immigrate to a cosmic "safe harbor" well before that certain collision, then we have no right to expect any other fate.

Therefore, I will not rule out the possibility that its perception is similar to our own perception of "self" and our bodies. If we can tell the difference between our thoughts and our lung-function, then it is clearly possible the Universal Mind can make the same distinction between its own mind and the components of its own "body" (galaxies, stars, planets, lifeforms, etc.).

If the Universal Mind can make such a distinction, then why wouldn't it cause evolution to create physiological adaptations to allow life forms within it to attain consciousness with the intent upon direct communication?

This communication would not be verbal. The universe is composed of many different star systems; each with its own, unique molecular disposition which will necessitate different physiological development. We can expect multiple life forms; each with their own unique sensory organs and appearance. Species around the cosmos will possess different languages as well. So, what unites us all?

Consciousness.

Consciousness speaks to consciousness without the need for language translation. Consciousness is automatic and instantaneous, just like entangled nonlocality. There is no need for translation practice, no need for innate talent or training in linguistics. The communication is consciousness to consciousness. There is no separation between source and receiver.

Loud sounds are heard through walls, because sound is a wave of energy that causes the air to vibrate. This vibration passes the sound along and can even pierce walls. Telepathy must operate on the same principle, but with photons.

Language is not the same as consciousness. Learning any spoken or written language requires training and practice. At its core, language is a nervous system process. You see words with your eyes. You hear words as sound waves through your ear canal.

Before your consciousness can become aware of either written or spoken words, your brain must first interpret the visual and auditory nerve impulses through your brain's vison and hearing centers. Then, these signals are routed to your prefrontal lobe for comprehension and appropriate response. You are not consciously aware of any images or sounds until they reach your prefrontal lobe.

The Universal Mind is conscious and self-aware by quantum entangled photons which exhibit quantum nonlocality. This is consciousness. All species of life, existing everywhere within the universe share consciousness because all life shares these quantum entangled photons and their shared quantum nonlocality.

Communication with the Universal Mind is inevitable. It is as inevitable as rain falling. All you have to do to "get wet" is to walk outside. But you have to make the effort. You must "go within" your own mind. You must dig down through your own consciousness with proper meditation over time to reach it.

It's as easy as opening a door. Does this remind you of anything?

"Behold, I stand at the door and knock. If anyone hears my voice and opens the door, I will come in and dine with them, and they with me."

Source: Rev. 3:20 Bible (NKJV)

The Universal Mind doesn't require religious faith or obedience to dogma. The Universal Mind doesn't require you to bow down or kneel down or lay yourself down to ANY Earthly authority. To reach the Universal Mind, the only "authority" you must overcome is your own ego.

Why do you think asceticism (the concept of self-denial) is so important to monks? Monks (regardless of religion) realize religion is not the key to spiritual salvation. For the monk, Earthly religion (even their own) is viewed as just one more material "thing" to be cast away in the search for true spiritual "salvation"—selfless enlightenment. Religions are created by humans and are, therefore, false gods and false idols. They are unworthy of worship. Spirituality is the only way to "look within."

True monks—whether they are Christian, Buddhist or something else—are alike in how they perceive spiritual salvation. For them, salvation is no more or less than mental clarity, which grows into serene enlightenment and growing oneness with the Universal Mind.

CHAPTER 19

Why We Die:
The Information Paradox

If you believe no species in the universe can be immortal, you are wrong. One such species is known even now to exist right here on Earth. The Turritopsis dohrnii is a small, transparent jellyfish that exists in all the oceans. This species of jellyfish avoids death by recreating itself into an earlier form of itself to avoid getting too old.

To be sure, they can be eaten by predators or killed by other means. However, if left to themselves, they can theoretically live forever. This is but one example of how evolution has been able to adapt a non-death genetic trait within a species.

"There are many other animal (and non-animal!) species that offer tantalizing glimpses into an ageless existence: the risk of dying for naked mole rats appears not to increase as they get older; the world's oldest known non-colonial animal, a remarkably stress-resistant ocean-dwelling quahog clam named Mind, only died (accidentally) after a good 500 years when researchers dredged it up out of the ocean and wanted to find out how old it was; incredibly ancient bristlecone pines seem to function just as smoothly as much younger trees do; a particular colony of quaking aspens is considered to be about 80,000 years old…

Do they hold the key to eternal youthfulness for humans, too? We know that ageing in humans is thanks to a multitude of factors, many of which we still don't entirely understand…"

Source: E. Berthold, "The animals that can live forever," Australian Academy of Science at science.org.au.

The Universal Mind did not intend for humans to die. Death is an evolutionary adaptation of the human body. Over billions of years of species adaptations and the development of genetic traits, evolution caused humans to experience death.

WHY? Why must we die?

You might say the answer is simple; that we die because our bodies wear out. But the issue is not as straight forward as it appears. If the universe is eternal, and stars exist for billions of years, why did evolution limit our life-span to only 100 or so years?

The answer is staring right at us. Let's put on our thinking caps and work it out together…

We must begin at the beginning, with universal laws. These fundamental laws create the limits for all existence. Not even the Universal Mind has the power to eliminate, nullify or ignore them. These laws developed out of the post-Big Bang plasma from which the present relationships of matter and energy were determined and set in motion. These fundamental laws are as "forever" as the universe itself.

One of these fundamental universal laws deals with information. We have discussed this law in great detail several times already. But let's go over this one law once more and see how it applies to evolution's adaptation of death for humans.

SECOND LAW OF THERMODYNAMICS (INFORMATION)

In all natural processes, the total information (the measurement of its change over time) of a closed system plus its surroundings cannot decrease. This means that the more time passes, the more information is created which cannot be destroyed. The universe creates increasing uncertainty and system unpredictability over time.

Now we add what the Information Paradox…

INFORMATION PARADOX

The human need for information (curiosity) is an evolutionary adaptation necessary for species survival and to create conscious contact with the Universal Mind. But as time passes, ever increasing information creates ever increasing uncertainty and system unpredictability—a threat to species survival.

Further, the instinct for species survival requires memories of negative events to be stronger than any other memories. However, too much negative information (memories) is harmful psychologically to individuals possessing increased self-awareness. This creates a threat to species survival on its own.

This Information Paradox has two separate causes. First, information is necessary for survival and for conscious contact with the Universal Mind. Second, negative memories are crucial to survival (Fall off a cliff? Remember to avoid cliffs), but also create mental illnesses which affect human behavior and, in so doing, threatens survival.

Evolution is not alive. It cannot "create" in the same sense we humans can. However, evolution does become altered by environmental change. Evolution does adapt to its environment.

Evolution found itself confronted with an environment in conflict with itself. Evolution was confronted by programs which were in direct conflict with one another. The solution evolution chose was as elegant as it was effective.

If human consciousness is purely linear, then it is just a straight line. With information, the line curves. This curve represents the relative change of species or individual behavior (change of state of the human system of consciousness). Too much information and too much memory closes the once-linear line into a closed circle. Consciousness is not enlightened.

This is just as unproductive an existence as the straight-line. A circle of consciousness is repetitive behavior that shows no growth. How do you become enlightened existing within a closed circle? You don't.

The solution is to transform the circle of consciousness into a spiral. Consciousness spirals upward when it experiences a degree of increased self-awareness. We call this enlightenment. It won't turn you into god or a devil—but it just might make you wiser in your next life than you are now.

Hence, death. Death creates a resolution to the Information Paradox. It is your life's "delete file" for that information your individuality no longer needs. The information your death leaves behind is not destroyed. It is simply "stored" forever in your past life.

Quantum mechanically, time is asymmetric. It does not run backward. Try as we might wish though we could, we can never go back in time. The "arrow of time" always points forward.

However, what we do in the past is never undone. It just remains in the past. Death ensures that the past is past and can what we experienced there can no longer be used against us—except for those memories which we take with us to preserve our unique individuality.

Most biological species on Earth die. But, as we discussed earlier, some species do not die. Death is an evolutionary adaptation—not a given.

CHAPTER 20

CERN and The Universal Mind

Located along the French-Swiss border near Geneva is a complex of tunnels and technology the likes of which have no match anywhere else on Earth. Officially located in neutral-Switzerland, the European Organization for Nuclear Research (CERN) is world's most sophisticated quantum mechanical laboratory.

This was not accomplished by accident. Established in 1954, CERN is supported by 23-member nations (all of which are European except Israel). Surprisingly, none of the major world powers (U.S. Russia or China) are allowed to be members. However, other nations around the world have since become associate members of CERN.

Its operations are very expensive. Despite financial woes throughout the continent, CERN's European partners spent over 1.2 billion Swiss francs to fund its research in 2021 alone. That is the equivalent of $1.314 billion US. Of that revenue, 927 million Swiss francs were spent on research programs and 487 million Swiss francs were spent solely on its accelerator programs. Furthermore, CERN budgeted 223 million Swiss francs in 2021 just to upgrade its Large Hadron Collider (LHC).

Source: Final Budget 2021, CERN at cds.cern.ch.

Why does CERN exist? On the day of this writing, CERN's own website acknowledged it was created to answer just two questions:

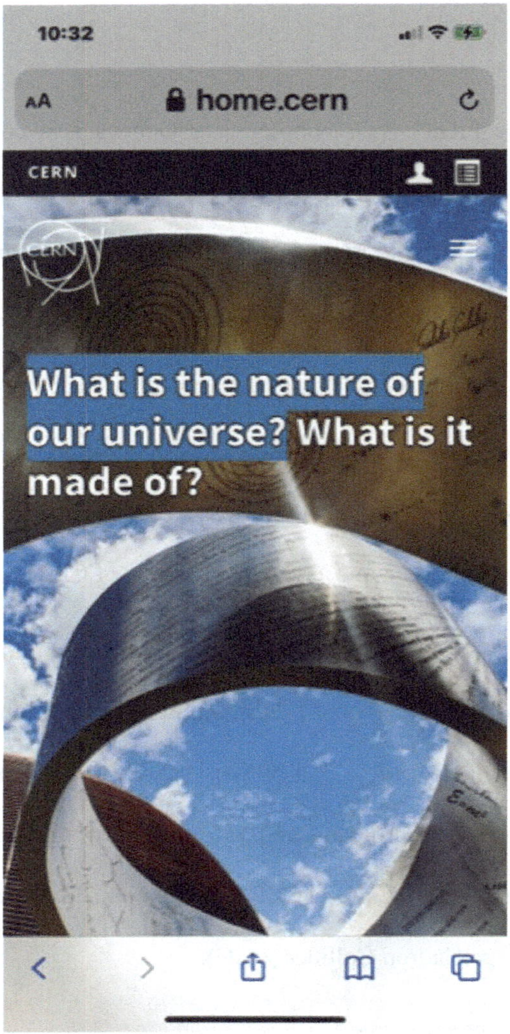

Photo: Courtesy of CERN. copyright 2022

Illustration No. 28. CERN webpage.

1. **What is the nature of our universe?**
2. **What is it made of?**

On the surface, these questions may appear at best benign and, at worse, only vague. But that is a deception. CERN may claim to be a "scientific research facility" but so does Los Alamos National Lab where the United States government designs and tests it latest nuclear weapons. Like most governmental authority and the United Nations, CERN wraps itself in a flag of symbolic imagery to avoid and deflect questions as to its real motivations and agenda.

The two questions CERN posts on its website are culpable admissions as to its charter—First Contact with the Universal Mind.

Illustration No. 29. The CERN dome in Meyrin, Switzerland. Courtesy: CERN copyright 2022

CERN is not new. CERN's inspiration originated in the 1940s when "a small number of visionary scientists in Europe and North America (Canada—not the U.S.) identified the need for Europe to have a world-class physics research facility. Their vision was both to stop the brain-drain to America that had begun during the Second World War, and to provide a force for unity in post-war Europe."

Source: "Where did it all begin," CERN.

These "visionaries" included Einstein's closest colleagues, physicists Niels Bohr and Louise de Broglie—each of whom contributed much to our present understanding of quantum mechanical theory.

It is not a coincidence that CERN was organized almost as soon as the Second World War ended. When Einstein proffered his theory on photon entanglement and nonlocality of the universe in 1935, no one really understood its implications. It was too much to grasp immediately.

But Hitler marched, world war raged, and physicists in America built nuclear bombs—proving once and for all time that quantum mechanics was a reality. These events left European physicists no alternative. If quantum mechanics were valid, so, too was Einstein's "spooky" universe. Bohr and de Broglie had no choice but to act. They both had come to terms with the majestic and awesome scope of Einstein's "spooky action at a distance":

1. **The universe exists as an independent sentient being.**
2. **Which Earthly power contacts the Universal Mind first may control it.**
3. **Whoever can control the Universal Mind may be able to weaponize it.**

These are the inevitable result of Einstein's understanding of how photons connect with each other and the rest of the universe in a single, entangled, nonlocal consciousness. CERN's members and associates are as dedicated to its purpose today is as they were in the late 1940s.

The quest for First Contact with the Universal Mind was tacitly acknowledged when India joined as an associate member in 2004. To celebrate the occasion, CERN accepted a gift from India's government: a two-meter-tall statute of the Indian deity, Lord Shiva. In accepting this gift, CERN acknowledged its creation is dedicated to finding the "life-force" of the universe. This can only mean one thing. The scientists and administrators at CERN understand their mission: First Contact with the Universal Mind.

"The significance of the Hindu God Nataraja (Nataraj) is said to be that Shiva is shown as the source of all movement within the cosmos, represented by the arch of flames. The purpose of the dance is to release men from illusion of the idea of "self" and of the physical world…The symbolism of Siva Nataraja is religion, art and science merged into one. In God's endless dance of creation, preservation, destruction and paired graces is hidden a deep understanding of our universe…Nataraja's dance is not just a symbol. It is taking place within each of us, at the atomic level."

Source: "Nataraja Hindu God of Dance, Destruction and Renewal," Lotus Sculpture, at lotussculpture.com (no date).

"In the Hindu religion, this form of the dancing Lord Shiva is known as the Nataraj and symbolizes Shakti, or 'life-force." Source: CERN.

Illustration No. 30. Lord Shiva statute at CERN. Courtesy CERN copyright 2022

Today, CERN operates ten particle-accelerators, decelerators and other nuclear facilities involving research into fundamental physics to "study" the beginning of the universe. The LHC is CERN's center-piece. Buried under 500 feet below the Swiss and French countryside in a long tube-shaped tunnel, the LHC spans over 27 kilometers in diameter.

Its purpose is to use massive magnets to focus sub-atomic particles into collisions to see what happens. There have been celebrated achievements, such as the discovery of the Higgs Boson in 2011-2013.

But CERN now plans to build a new, much larger super-collider—the Future Circular Collider (FCC). Even now, CERN is seeking funding for this new, and greatly expanded project which will build a super-collider over 100 kilometers in diameter. The extra distance in acceleration is needed to advance particles to as close to the speed of light as possible.

Why is this needed?

CERN was not created merely to study quantum phenomena and try to prove the existence of gravity waves or the Higgs Boson. CERN was not built to do "ordinary" science. CERN's LHC and intended FCC are communication devices. They are worm-hole creators with but one purpose—First Contact with the Universal Mind.

To hack into a computer program, even a supercomputer built on classical mechanical principles, one still must find a way "inside." Hacking into a human-designed computer running a program via an electrical processer conforming to classical mechanical physics, all you need to do is gain control of an electron.

If you want to "hack" into the Universal Mind, you need to find a way to open a channel in a quantum mechanical field. The only way you can do that is through controlling photons. But before you can control a photon, you must first "catch" one—you must "catch-up" to its wave-signal speed (light speed)—or get as close as the laws of quantum mechanics allow.

This is precisely why CERN needs a much larger accelerator, the FCC. Before they can "crack and hack" an entangled photon to communicate with the Universal Mind, CERN first must catch one.

But all they have to do is catch one…just one. As all photons are entangled and all share nonlocal continuity with the Universal Mind, each photon is like a universal cell phone.

What do they intend to do when they "crack" this hack? Does CERN spend the last seventy years and billions of dollars just to walk up to a cosmic microphone only to say,

"Hello?"

Why settle for just First Contact? This is not some minor league event. This is direct communication with an infinite consciousness in control of EVERYTHING. Whoever gets First Contact will demand SOLE contact. CERN's funding partner-nations will seek to establish an on-going relationship with the entity for sole control.

Once sole control is attained, will weaponization be next? Why wouldn't it? When was the last time humanity discovered a force of nature it DIDN'T weaponize? What could stop them from trying?

Furthermore, despite CERN's promises that all its research will be made public, do not expect its contact with the Universal Mind to be revealed--ever. While CERN's original charter promises to make all scientific discoveries public, the discovery of the Higgs Boson cannot be compared with First Contact with the Universal Mind.

In the European Union's own self-interest, CERN probably long ago was ordered by its governing partner nations that such "extraordinary" research as First Contact is beyond the scope of its original charter of purely "scientific discovery" and that all communications with the Universal Mind shall remain secret.

CERN has no choice but to keep such contact ultra-secret. Any attempt at First Contact would implicitly confirm the Universal Mind existed and that an international coalition was intending to gain access before anyone else.

Disclosure that CERN was even attempting to make First Contact with the Universal Mind would create an international furor that would likely result in the use of force. It would certainly test the limits of Swiss neutrality-protections from foreign interference and may result in war.

Both Iran and the Taliban, who now rule Afghanistan, both adhere to Sharia Law, which was the ultra-violent philosophy of ISIS. Can you imagine what the radical terrorists who rule Iran and the Taliban would do if they learned an attempt at First Contact was even being considered?

Both Iran and the Taliban would begin nonstop terrorist atrocities; both in Europe and around the world. The European Union would then have to make a choice: attack Iran and the Taliban on their own or seek American military assistance through NATO.

Can you imagine a US President having to consider a SECOND invasion of Afghanistan AND attacking Iran at the same time?

What about the world's other dictatorships: China, Russia and North Korea? How will they react to the possibility of Europe gaining First Contact? Will CERN's "official" residence in traditionally-neutral Swiss soil be enough for them to refrain from a military attack of their own?

China and North Korea are both extremely anti-religion. Each sees the practice of ANY religion by an individual as an attack against the state resulting in severe, even capital punishment. Since the communist take-over of China, Chinese communists have imprisoned, tortured, and murdered untold thousands of Buddhist monks and populace. Hundreds of Tibetan Buddhist temples and monasteries have been destroyed—all because the totalitarian, thuggish communist regime known as the "People's Republic of China" is anything BUT about freedom of people or a democratic republic.

China and its client state, North Korea, imprison, torture and murder dissidents. Free speech does not exist in either country. Both dictatorships would see the Universal Mind as an enemy of the state. Would their paranoia take them to war?

Sources: "Human Rights in Tibet," freeTibet at freetibet.org. and "Early History; The Invasion of Tibet; Present Situation;" Official Students for a Free Tibet website at umass.edu.

Russia is no longer communist. It's government supports Orthodox Christianity once more as ex-communist KGB officer Putin attempts to recreate the Holy Roman Empire in Eastern Europe. Putin is a pragmatic dictator in the Stalin mold. He invaded the Ukraine to regain the buffer zone of satellite states that insulated Mother Russia during the days of the Soviet Union.

He suppresses dissent and assassinates his foes relentlessly. Neither Putin, nor any other Russian leader who succeeds him, will accept an independent Europe's claim to sole contact with the Universal Mind. Will they conclude the stakes are too high to allow that to happen?

But CERN likely wants more than First Contact. It will want SOLE contact. If CERN and the European Union can keep their project secret, they are likely to get it. No other group of nations has funded such project. CERN will control all contact going into and coming from all communications with the Universal Mind.

In fact, First Contact may have already happened, and CERN has kept it secret from the rest of the world. They may have discovered that communication with the Universal Mind is hampered because the particle-collisions used to "open a channel" are too slow to achieve the level and quality of communication they feel they need.

Why else would CERN suddenly demand funding for the FCC? Why else is there any need to get to achieve light-speed particle acceleration? Let's take a look at just how BIG their dream super-collider will be…

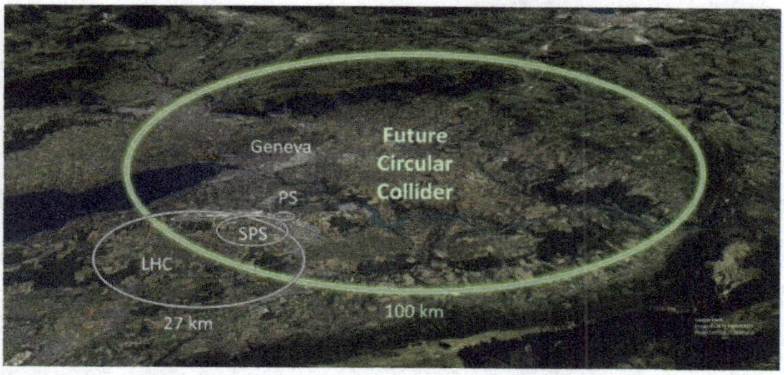

Illustration No. 31. 27 km. LHC & 100 km. Future Circular Collider.

Courtesy: CERN copyright 2022

In the above photo, the Large Hadron Collider presently being operated by CERN is shown in the bottom left (LHC), intersecting other, smaller accelerators SPS and PS now being used by CERN for experimental research unrelated to "photon-hacking."

The above photo demonstrates the enormity of the yet-to-be-built Future Circular Collider. The FCC is an upgrade of staggering proportions and makes the present LHC-collider appear miniscule by comparison.

Yet this increase is needed to achieve the particle-speed necessary for CERN to be able to finally "crack the hack" of a photon. If CERN is successful, it may become THE most significant event in human history—First Contact with the Universal Mind.

And there are few, if any, options any non-member nation have to stop it. It is funded solely by its members. Buried under 500 feet of French-Swiss soil, CERN is protected from military force. Not even a nuclear attack would harm its staff or equipment or disrupt its operations.

It is protected politically by Switzerland's traditional neutrality. Swiss neutrality dates back all the way to 1515 following its army's devastating loss to the French at the Battle of Marignano and was permanently made Swiss foreign policy during Napoleonic times.

Source: E. Andrews, "Why is Switzerland a neutral country?," History at history.com.

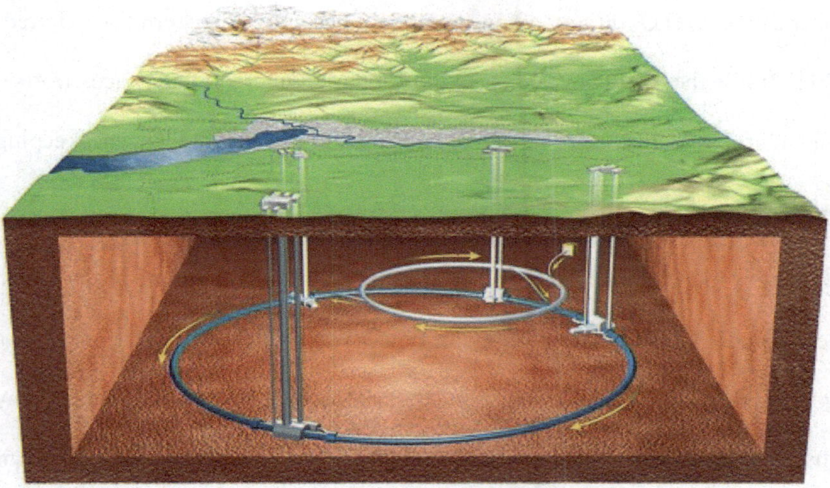

This Photo by Unknown Author is licensed under CC BY-NC-ND

Illustration No. 32. Underground CERN instruments and staff are safe from even a nuclear attack. Courtesy: CERN copyright 2022

CERN'S member nations determined that its facilities and research would be protected from both political and military interference—especially by the U.S., Russia and China. That is why all of CERN's technology is buried in nuke-proof tunnels extending between France and Switzerland.

CERN is anti-military as well as anti-world-power. CERN's originating convention states:

"The Organization shall have no concern with work for military requirements and the results of experimental and theoretical research shall be published or otherwise made generally available."

Source: CERN

The Swiss army is more a domestic militia than a military force. Switzerland is not a member of the NATO military alliance and only became a member of the United Nations in 2002. While the Swiss have consistently supported international peace efforts since the Second World War, they have repeatedly declined to join actual UN peacekeeping forces.

The Swiss may appear weak militarily. CERN may appear to be at risk. But that appearance is misleading. There is a reasonable likelihood that none of the major powers will ever attack CERN because no major power wants to destroy their one, major source for gold production. CERN's single-greatest defense against military attack may just be because most nations and many businesses and private individuals keep their money in Swiss bank accounts.

CERN's most valuable protection is the fact that Swiss bankers have possession of billions of dollars of national and private gold, silver and paper currency. Switzerland is not only the world's banker, but the world's most accomplished gold refiner. Today, most of the gold in the world passes through the experienced hands of Swiss gold producers.

"This is a business worth CHF 70-90 billion ($70-90 billion U.S. dollars) depending on the year. Gold arrives here in unrefined form and leaves the country in all its glittering purity…Swiss refineries process 70% of the unrefined gold mined in the world each year. Four of the nine major players in the global gold industry conduct most of their business here in Switzerland. While the gold originates in ninety different countries, the United States, Russia and China all send their unrefined gold to Switzerland for refining."

Source: R. Franciolli, "The shady origins of gold refined in Switzerland," Swissinfo at swissinfo.ch. 2019.

Swiss banking is also famous for its stability, confidentiality, and privacy, as well as its famous national neutrality:

"Swiss privacy laws include strong penalties for any bank or individual who divulges client information to any third party. Additionally, Swiss courts do not recognize foreign judgements and will not lift secrecy laws for civil matters such as litigation or divorce cases. Needless to say, Switzerland has rightly earned a reputation as one of the world's most private locations."

Source: Harbor Financial Services, "Swiss Bank Accounts," at hfsoffshore.com (no date).

U.S. foreign policy interests are fundamentally in conflict with both their Russian and Chinese adversaries. But all three major powers agree on this—they ALL need the Swiss to remain independent. Everyone needs a banker they can trust.

CERN should be safe from both political and scientific interference as well as direct and indirect military attack. CERN's quest for First Contact may go largely unchallenged and unpublished.

Has CERN already made First Contact? It took decades and billions of Euros to build the 27-kilometer LHC collider. Suddenly it isn't good enough. Suddenly it is not sufficient to their needs.

WHY?

Have CERN's scientists already made First Contact with the Universal Mind?

What other conclusion can logically be drawn from CERN's sudden plan for a 100-kilometer Future Circular Collider. CERN must have a reason to believe that its present 27-kilometer Large Hadron Collider (LHC) is insufficient.

For what possible purpose is the HLC not good enough? What possible purpose could a 27-kilometer accelerator that took decades and billions of dollars to plan, build and test be insufficient to accomplish? The HLC was powerful enough to detect the Higgs Boson. Surely, it is powerful enough to do any other quantum mechanical science. But, apparently, CERN disagrees.

Why?

This sudden, inexplicable "need" for an accelerator almost four-times more powerful than the one CERN has is all the proof we need to know "scientific research" is not the real reason CERN exists. Just as the real purpose of the Los Alamos National Lab in New Mexico is not academic science but nuclear warfare research and development for the United States government. CERN's real purpose is now just as clear.

First Contact.

The science CERN does is just a cover story to justify the incredible sums sought to build and operate CERN's main project—First Contact with the Universal Mind. It is now obvious the LHC and FCC accelerators are not merely scientific instruments, but sophisticated communication devices using entangled photons to create an artificial "nonlocality" for but one purpose—communication with the Universal Mind.

Assuming CERN has already attempted limited First Contact, what was the result? Since CERN will never disclose such events voluntarily, we have no choice but to speculate.

What would CERN choose as its initial message? How do you introduce yourself to your God?

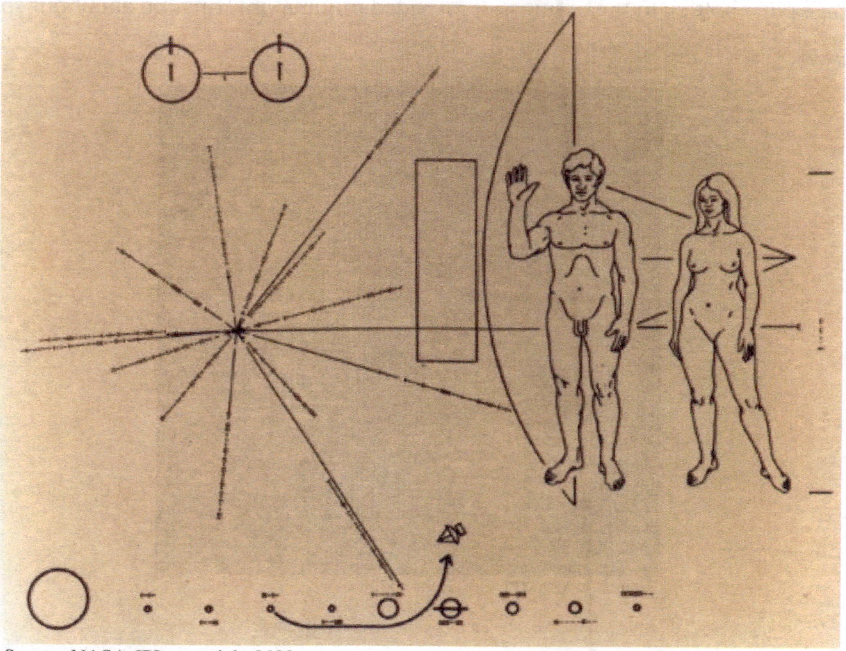

Source: NASA-JPL copyright 2022

Illustration No. 33. The Pioneer spacecraft plaques.

The illustration above shows the gold-anodized aluminum plaques placed on board both Pioneer 10 and 11 spacecraft. The Pioneer probes were launched in 1972 as Earth's first "messages in a bottle" to the cosmos. It provided minimal information about our species.

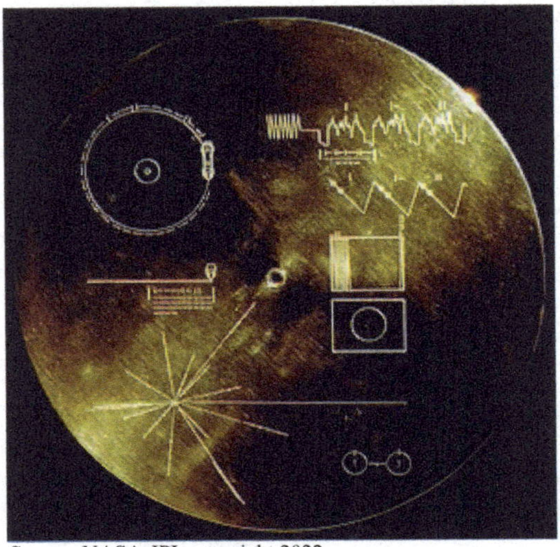

Source: NASA-JPL copyright 2022

Illustration No. 34. The Voyager Golden Record.

In 1977, NASA launched Voyager 1 and 2. Directed toward the far reaches of interstellar space, each probe carried with it what has been called a "Golden Record." The Golden Record is a 12-inch-diameter disc of gold-plated copper containing much more information about Earth and our species than the simple plaque on board the Pioneer craft.

The disc's contents were chosen by a committee headed by world-famous astrophysicist and cosmologist, Carl Sagan of Cornell University and included over a hundred images and a broad selection of natural sounds, music and spoken greetings in fifty-five languages.

Source: JPL-Caltech

We should take a moment to note the unfortunate distinction between NASA-JPL's attempts at interstellar contact and CERN's. All four Pioneer and Voyager probes were planned, prepared, launched, and controlled post-launch with full disclosure to the rest of the world. Nothing was kept secret from the rest of the world. Dr. Sagan and his colleagues pursued an inclusive approach toward contact with alien species. Even the messages carried by each probe were publicly disclosed well in advance of launch.

CERN and its partner-nations have apparently made a different choice. Once the world understands the methodology and application of its LHC and FCC accelerators, CERN will inevitably draw wide-spread criticism for their secrecy and non-inclusivity.

While this criticism is warranted, it will change nothing. The European community funding CERN will never surrender its advantage in the race for First Contact.

The Pioneer and Voyager probes and their simple messages give us something to work with as to how humanity will introduce itself to its god through CERN. But contact with the Universal Mind is not a function of a probe into outer space. First Contact with the universal consciousness will be via inner space.

Let's look into the future…

Somewhere, five-hundred feet beneath the CERN dome, is a vast room. Environmentally controlled within very strict guidelines, this room houses sophisticated instruments: ten quantum computers linked in series with each other through quantum entangled photons; data from over 100,000 tape-cartridges (each with a memory capacity from 1 terabyte to 5.5 terabytes) are stored within CERN's Advanced Storage (CASTOR) system; and memory storage capacity over 1000 petabytes to handle the almost infinite amount of information expected to be received from the Universal Mind.

Separated from this equipment, inside a perfect anechoic chamber, sits a woman wearing a cap. She is an Emissary; one of a select-few humans given the honor to serve as human representatives to the Universal Mind. The cap is connected electronically to an implant that allow her to communicate from her mind directly with the Universal Mind.

This cap is the direct interface between the Emissary and the Universal Mind. Everything her consciousness experiences, from telepathic thought to images and sounds she receives from the Universal Mind are stored in CERN's vast CASTOR memory network.

Only young women are chosen as Emissaries. Males are not permitted to be Emissaries. There are several reasons for this.

The role of Emissary will be an extraordinarily stressful one psychologically. Contact with the Universal Mind is unpredictable and subject to extremes; due both to the intensity of information communicated and the emotional responses evoked from images received from the Universal Mind.

There will be physical pain, as the Emissary attempts to be a living window-pane, absorbing infinite data at light speed through her finite brain capacity and on into CERN's vast memory storage center. It feels like an acid trip accelerated a thousand times.

Imagine stepping into a column of pure intelligence; blindingly white light sears your mind down to its individual neurons. You perceive the universe. Infinity stares at you as you begin to contemplate the meaning of fundamental natural laws. You begin to comprehend space and time…then… the connection ends.

In a nanosecond your mind is emptied of all you just experienced.

How many times can you physiologically and psychologically adapt to that before you break? No one knows. But we must accept the fact that interaction with the Universal Mind will take its toll on the chosen Emissaries—both physically and psychologically.

The American Psychological Association studied how each gender deals with extreme stress (physical and psychological). Their findings are revealing. It appears that whatever advantages males have in physical strength are significantly outweighed by their emotional immaturity. Males are less likely to admit to weakness and unwilling to admit they need help.

"Likewise, men put less emphasis on the need to manage their stress than women do. Men see psychologists as less helpful and are less likely to employ strategies to make lifestyle and behavior changes."

Source: (no author), "Stress and Gender," American Psychological Association, apa.org.

This is not surprising. Generally-speaking, male decision-making is often compromised by their testosterone-loaded egos. As a result, they are generally more likely to refuse behavior management techniques to reduce the stress of repeated and intense communication with the Universal Mind.

This ingrained (genetic?) self-reliance makes males unacceptable for the Emissary program. Males are basically unreliable and uncontrollable. Their self-awareness is limited and is often overcome by pride and ambition.

Furthermore, the most important psychological skills an Emissary will need are adaptability and selflessness. These are traits which women have in abundance. Men do not.

Merging with the Universal Mind will test one's ability to lose oneself in an infinite-consciousness. Males will fail. Humanity cannot risk failure.

Researchers have found that women's brains are built differently than male brains, allowing women to be naturally more self-less and able to disconnect ego from decision-making.

The "neural reward system appears to more sensitive to prosocial (selfless) rewards in women than in men, providing a neurobiological account for why women often behave more prosocially (selfless) than men."

Source: A. Soutschek, C. Burke, et al, "The dopaminergic reward system underpins gender differences in social preferences," Nature & Human Behavior, Oct. 2017.

Given the historical significance of such a First Contact event, males are more likely to succumb to personal ambition at the expense of the Emissary mission. Personal glory, financial wealth and fame would likely be substantial temptations for anyone involved. Males are more susceptible to these corrupt influences than females. Furthermore, males are more likely to resist management intervention when needed and resist commands at the risk of the mission.

Females, on the other hand, are more self-less and willing to cooperate with command decisions. Males are too psychologically fragile, too emotionally unstable to risk being used as Emissaries. But they make great guards.

Psychological screening of recruits is necessary to ensure each Emissary selected meets the standards required for interaction with the Universal Mind as possible. Each of the CERN partner-nations are responsible for recruiting prospective Emissaries. These recruits are evaluated using a cover-story that they are candidates for a top-secret intelligence program being planned by their respective nation's military.

Recruits who pass this initial phase are sent to a private hospital serving as a cover for the second-phase which includes more sophisticated psychological screening and extensive training. This phase will take several years and will require its own separate organizational and training structure, separate from CERN.

It will be from this pool applicants that the novice-Emissaries will be selected. Each novice-Emissary will make the journey to CERN, under cover. She, and the others like her, will travel as part of tourist groups, disappearing into CERN's secret facilities without anyone knowing what happened to them.

Still more training and psychological testing awaits the novice-Emissaries upon their arrival at CERN. Only when these final tests are completed does each novice-Emissary come before The Committee; a select-group of CERN researchers, scientists, and administrators given the responsibility of choosing which novices graduate to full Emissary status.

As everything involved with First Contact must be totally secret, the consequences of such secrecy will inevitably govern much of the Emissary-selection criteria. For example, here are a list of such criteria necessary to limit the risk of program disclosure to the rest of the world:

1. The Emissaries must commit to living in the CERN underground indefinitely. A term of one-year can be attempted, but what happens if the Universal Mind gets obsessed about one of the Emissaries and doesn't want her "retired?" A shorter-term of service should be considered to prevent the Universal Mind from becoming too codependent upon any one Emissary.

2. Emissary-candidates will have to be single, with no children. Even parental and indirect family relationships may be a reason to reject someone as the Emissary's consciousness must be as clear as possible when in communion with the Universal Mind.

3. Emissaries will need to be trained to keep their minds open as long as possible. Their role will not be for their own personal communication but solely as a lens to receive whatever the Universal Mind sends to CERN and vice-versa.

4. Provisions will need to be made to house and secure the Emissaries once they are "retired." CERN security will depend upon keeping these former Emissaries from being kidnapped by foreign governments or harassed by media. The only way this is practical will be for CERN's partner nations to build a community in which former Emissaries and CERN staff can live above-ground in total secrecy indefinitely.

Why use humans as Emissaries? Why not use quantum computers as the interface between humanity and the Universal Mind? There is a very simple answer to this. Do you really want a computer you cannot control as your link to the Universal Mind?

The Universal Mind works on the same quantum principles as the quantum computer you are using for communication and memory storage. They are both sharing the same entangled photons.

How do you trust a quantum mechanical computer interfacing with such a highly advanced intelligence communicating privately through each other's shared entangled photon link? How do you know the Universal mind wouldn't take control of your own computer? Would you be able to tell if it did?

But there is an even more compelling reason for using humans as the interface with the Universal Mind rather than a computer or artificial intelligence.

First Contact with the Universal Mind will be the greatest event in human history. Relying upon machines to do our communication for us would be worse than embarrassing—it would demonstrate our unworthiness as an intelligent life form. For all we know, the Universal Mind might just terminate us right then and there. When an important guest asks for the master of the house, don't send the butler.

If you are still unconvinced that CERN's LHC particle accelerator and its planned FCC accelerator are actually communication devices intended for contact with the Universal Mind, consider one more piece of evidence.

Back in Chapter 15, we discussed how in 1964 John Stewart Bell proved Einstein's theory of quantum nonlocality (spooky action) was valid. Bell even wrote a famous equation (Bell's Theorem) explaining how nonlocality works. Do you recall from Chapter 15 who employed him when he made those incredibly significant discoveries in 1964?

CERN.

CHAPTER 21

"Resurrectio et vita nova..."

The change is sudden. One moment you are struggling to take a breath, your chest heaving. The next moment and the weight vanishes like the mist in the morning sun.

You realize why.

You no longer are breathing. An autonomic impulse you have taken for granted your entire life is no longer functioning. Expecting to fall unconscious, you are surprised when you remain quite alert. To the contrary, you find yourself more aware mentally than you can ever remember.

How can this be?

As your physical body begins to shut down, your consciousness reaches a higher focus. It is an odd sensation, not having to breathe; one you find exhilarating. Its liberation presages the change to come. Your mind races, desperate to understand what is happening.

Part of you is still stuck on how weird it is to be awake without breathing. Another part is slowly coming to a realization—one you resist.

You try to move. "Stand up!" You tell yourself. "Shake it off, get on with it!"

But nothing changes. Nothing changes at all.

Panic rises like a voiceless scream within you. Desperately, you try to choke it back. Am I dying?

You feel a gradual withdrawal of sensation within your body. The sense of touch is the first to go. The pain is gone.

But so is everything else. Voices that were once clear are now muffled; the pitch of their voices gradually lowering and becoming slower until you can hear them no more.

Your last memory of sight; was it a face? Whose face was it? You are no longer sure who it was you saw.

Your entire life flashes across your mind like a motion picture. It takes less than a second. Your mind is gradually separating from your dying physical body.

The brain being the last organ to die, you still retain enough contact with your short-term memories and emotions to picture your loved ones in your mind. A deep sadness engulfs you as you realize you will leave them behind.

Your mind screams, "I don't want to go!"

You suddenly feel a "presence" within your mind.

"Come…come with us," a voice in your mind beckons. It is not your own. Nearby a portal opens, a white light shines within. "You are no longer needed here. It is time."

As you float toward the portal, most of your memories evaporate like steamed water. They are no longer necessary. Your most important memories from your life are stored within the pool of memories from all your past lives. But now that your mind is connected to all your past lives, you realize that the life you just left is no more important than any of the others. For all their differences in birth, life-style, health and longevity, they all share one important trait—they are all past.

As you enter the portal, you see your own reflection, a tightly wound glob of intermingled photons which could be mistaken for the glow of a light-bulb. Within that glowing ball of photons are all the memories that make you the individual you have become over time.

But you also see something you did not expect. Escorting you are two other glowing photon-orbs; your Guides for your journey to the "other" side of existence. You "hear" one of them give you a telepathic instruction:

"Think music…"

Your mind searches for something musical. Out of the blue, you remember an old rock and roll song. You fill your mind with its melody as your photon-being is stretched into an elongated wave-form as you travel at light-speed through a worm-hole. Your wave-form, like all other light-waves has both a peak and a trough. You have your own unique frequency and wave-length.

"Everyone's wave looks different," your Guide observes. "It will determine many of the characteristics of your future life. Individuality requires continuity from life to life. As you can see, ours is very different from your own."

It is true.

Your Guides' wave-forms are much more compact and taller than your own.

The worm-hole is very large. It needs to be. Inside, there are many groups of wave-forms just like your own. They appear to be coming from all directions, all heading toward a single point in space-time. Each consists of two guides with wave-forms that are compact and tall escorting a wave-form that is longer and much shorter.

"Where are you taking me?" You ask.

"Home."

Growing in confidence, you press on for more information…

"Do you have a name?"

Your Guides laugh.

"Names are for shadows," your Guide responds. "Where we dwell, there is no separation, only knowing. I can think of you and instantly know everything you have experienced in all your past lives that has made you the individual you are now. Isn't that better than a stupid little name?"

"What are these 'shadows'?" You ask.

"You, of course." The Guide answers. "You and your kind are but shadows to us."

"Why?" You ask in amazement.

"You and your kind dwell in a squalor of self-ignorance," your Guide responds. "Therefore, your wave-forms are more physical than light. You are not yet sufficiently self-aware. As a result, you remain physical, impure. That is why you and your kind remain bound to the physical realm. Where you dwell, your physical bodies block the light. Your existence creates shadows where light should be."

Your journey through the worm-hole ends as suddenly as it began. The Guides immediately place your photon-orb into a cocoon-like shell.

"This shell will keep your wave-form stable until we are finished." You are told.

That frightens you.

"What do you mean, 'until we are finished?'" You ask. "What are you going to do to me?"

"Do not be afraid, little shadow," your guide's thoughts attempt a soothing tone. One you are no longer certain you can trust. Everything is so strange.

"Your wave-form cannot survive here for long without protection," the guide explains. "That is why we placed you in a protective-shell. It will keep your individuality stable through your redeeming and implantation."

There is no point to resisting. You are dependent upon your guides for survival in this alien environment. You stop questioning. You look around.

You and your Guides have joined a long queue. Millions of photon-orbs just like yours are slowly making their way along a procession-like pathway. This place has no official name that I am aware of. However, I call it "ShadowLand."

At the end of the pathway is a large temple with four pillars. In front of the temple, on the left side, is an altar surrounded by the same guide-like beings who attend to you. On the far other side, a pillar of incandescent, white light stands very much alone.

Overhead, shadows float like clouds. They descend, passing by like a dark fog. You are in a land of light and shadow, where matter and energy share space and time as one.

You reach the end of the queue. One of your guides examines you closely.

"Your attention will be required shortly," the Guide warns. "There are…formalities to observe before your rebirth."

"You're sending me back?!"

"Yes," the Guide answers. "Your wave-form lacks the focus to advance to the next realm. Your consciousness remains bound to Earth. I am sorry. You must return."

The shell containing your photon-orb floats along as your Guides escort you to what appears to be an altar of some sort. Three figures who look like Guides float in front of you, awaiting your approach. The tallest stands in front. The other two, shorter, stand respectfully, behind. As your shell-pod reaches the altar, the tallest one floats over to you and bows.

Your shell-pod opens. The taller of the three examines you. A focused beam of light is directed at you. You are scanned thoroughly.

"I am Tourine, of the Priest-Clan Mamet. Your soul is healthy and ready for the Resurrection and the New Life."

Tourine leads you out of your shell-pod. Your photon-orb floats freely now.

"Come," instructs Tourine. "You have some choices to make about your next life."

"What kind of choices?"

Tourine instructs you to silent attention. You are led to the altar. Seated upon the altar are four golden chalices. They glimmer in the light, reflecting the light brilliantly around you.

Each chalice has upon it a face—a human face. Two are male. Two are female. Tourine fills each chalice with what looks like red wine and replaces them upon the altar.

Tourine turns back toward you. It stands before you and spreads its wings(?)…arms(?). It's "voice" becomes amplified, grander, ceremonial in tone.

"Your future life stands before you in one of these four chalices," Tourine's voice fills your consciousness. "They have been prepared for you and you alone. You may touch each chalice for a brief vision of what each life would hold for you. Once you are finished, you must choose one of them and drink."

Your body of photon energy moves toward the altar. As you reach out to touch to the first chalice, you finally see yourself as you are now, in this alien realm. Your arm is a warm yellow plasma, with deep-seated shadows floating throughout. The shadows sadden you. They are what is keeping you from moving on to whatever lies beyond the Earth. For a brief moment, you wonder if you will ever be able to find out.

Your plasma-filament appendages touch the first chalice…

What you have just read is based upon what I experienced while I was in this realm before being reborn in my present life. Except for close loved ones and a very few, select friends, I have kept this hidden for all my present life.

The time spent in this realm is part ceremony and part practical application. It is your one opportunity to choose which life you will live when you are reborn. Touching each of the offered chalices affords you a glimpse of the life you would live if you chose that particular chalice.

This "glimpse" is not make-believe. The "Guides" prepare each chalice just for you. Each chalice contains a moment in real time which you will experience should you choose the life contained in that chalice. It is a real moment you will live in that particular life. The "glimpse" will be short, no more than one incident you would live through. But each such "glimpse" will be enough to know who and what you are as a person in that life.

You may be any race, any nationality, any gender, any sexual preference; with one provision. Your past lives have several effects upon your future lives. First, your past life choices create a center of gravity which preordains your future life options.

In other words, what race, gender, nationality, sexual preference you choose in past lives will determine to a significant degree what options are made available to you for the life to come. These choices do affect your abilities to excel in your next life. For example, if you have lived many lives in English-speaking countries, your language skills will be much better being reborn in an English-speaking country again.

If you choose a non-English-speaking country to be reborn in, you will find learning that language difficult and your social skills will suffer as a result. The result is that most people usually prefer to be reborn in the nation they have lived in before—or at least they keep to their same language.

Also, while your physical characteristics are determined by the genetic lineage of the body your will be reborn into, you retain your intelligence level and your hand-eye coordination—good or bad. Why do we retain intelligence and hand-eye coordination from life to life?

Remember, intelligence is the evolutionary ability to process information. Information can never be deleted. Therefore, you never lose what intelligence level you have. You can never be born with less intelligence than you ended the last life with. You always enjoy the potential to increase your intelligence level from life to life, however.

Hand-eye coordination is a function of intelligence. Reflexes are intelligence-dependent because they are a function of information processing which we carry from life to life.

Karma also plays a significant role in what choices you are offered with each chalice. Every chalice offered to you will contain Karma-causation life events within it.

There is a problem, however. The vision you receive from each chalice will not reveal how your Karma-history will impact your next life. Your specific Karma is hidden from you.

Think about it. Would you accept a future life if the vision a presented by a chalice showed you a Karma-future you did not like? I doubt any of us would.

My memory of ShadowLand is quite vivid. I did not trust my ShadowLand Guides. I brought this on myself.

My past lives were full of violence. Too many lives lived as a warrior, mercenary, or assassin. I can recall but one past-life where I enjoyed a long-life. I cannot remember any where I had a family of my own. I usually died in battle, young, fearless and stupid or middle-aged and worn out by years of marching and battles fought for rulers who cared not what happened to me or my brothers-in-arms.

I feared my past-life Karma would haunt me in this life. When none of the chalices offered me revealed any Karma, I rejected them all. Believe it or not, there is an alternative to the chalice-choices. The following is as close to what the Guides told me about this alternative as I can recall:

When I refused all the offered chalices, a new Guide floated toward me. This Guide was taller than the others and moved with an energy of authority and power I had not perceived from any of the others.

"I am Tourine, the Chief Priest. Look over there," its "arm" gestured toward the far-right side of the altar. There, standing alone, was what appeared to be a vertical beam of pure energy. Shimmering as though alive with energy, it stood mute. A pulsating plasma of unknown power.

Was it peaceful or malevolent? It was impossible to tell. But none of the Guides went near it.

"We call it the Pillar of Light." Tourine's thoughts betrayed a reluctant reverence. "If you refuse the chalices, the only way to be reborn is to merge with the Light."

I wasn't sure that was my best option either.

"If I go into the Light, what will it do to me?" I asked.

"The Pillar of Light does what it wills. It does not take requests."

"What is it?" I ask. "Where did it come from?"

"None of us know." Tourine admitted. "It was here, waiting for us, when we arrived. All it does is what you see. It stands and watches. What is it? We can only speculate for very few have dared enter into it. Most who have do not come out. And those few who do, return much different than when they entered. We leave it alone. You would be wise to do the same. Choose a chalice and drink."

"No," I replied. "I choose the Pillar of Light."

"So be it," Tourine waves an arm. The chalices disappear. I am led to the Pillar of Light.

Stepping into the Pillar of Light was like overdosing on acid, but a hundred times greater. It was taking a shower in an infinite sea of consciousness. It was heaven…and it was hell; like tripping on acid, but a thousand times more intense.

An infinite library of cosmic knowledge warped through my mind at the speed of light. Images of natural laws, insights to predict human behavior and events, visions of planets with life that looked surprisingly like ourselves; and visions of our own future here on Earth; all revealed in what must have been only a microsecond.

Screams transformed into giggles of delight. Fear replaced by insight. Everything will be all right…eventually. For a brief moment in time, I saw the universe through the eyes of the Universal Mind.

As soon as it began, it was over. I found myself floating outside the Pillar of Light. The beautiful knowledge, the sublime insight, all the eloquent formulae, the infinite well of information; it all slowly faded away. The human brain cannot retain such a vast amount of information.

CHAPTER 22

Mind Over Matter

I know I am not the only one whose life experiences have taken them beyond the frail fantasy of their myth-made religion. There are many who have questions and who struggle with the search for answers. I hope the following helps…

Who Are the Guides?

They are not angels in the religious sense. None had wings. The do not represent any god. During my Near-Death-Experience among them as a child, their communication with me was mind-to-mind.

They appeared and left as a group. They all appeared the same to me, what you might describe as coherent plasma orbs that possess some sort of active energy. It is not a pulsation. Nor did they radiate. It was more of an ionization effect, like how a ball of lightning might look like if it were stationary.

Not once did the Guides mention Yeshua, Jesus, any saint or prophet. There was no mention of any religion. Not once did they acknowledge any Earthly religious concept or myth at all. They never used the words "heaven" or "hell."

I refer to them as "Guides" because their actions were always to assist me in coping with the experience of dying or aiding my processing to ShadowLand. They constantly advised me they would provide me with a "better body" if only I would leave with them—which meant leaving my family behind.

The Guides did not tell me what I had to do. They let me decide. When I chose not to leave with them, they did not argue with me. Instead, they left in as orderly a manner as they arrived.

Their behavior reminds me of how we attorneys interact with our clients. As an attorney, my responsibility is to advise and counsel my clients on the legal process they must go through, the choices available to them and the consequences of each choice they may make. The decisions on which option to choose are always up to the clients.

The Guides do much the same for our souls. They educate us, make us aware of certain options in our lives and their consequences. The right to choose remains with us. The Guides never forced anything on me.

I still recall their presence at the hospital and have drawn some tentative conclusions from my observations of them. The way the entered and exited the trauma room (through a portal of some kind through the ceiling) had the look of a formalized procession. About ten entered together and remained near the portal. Two other orbs then emerged through the portal and floated down to me. The original ten remained in place as though they were witnesses or were subordinate to the two which came to me.

Therefore, it appears they obey an organized hierarchy of some kind. I also detected definite male and female genders in the inflection of their telepathic communication with me. This leads me to the conclusion that if one chooses to be one gender over the course of enough lives, one's unique individuality (soul) begins to take on the traits of that gender over those of the other.

Guides have been associated with angels in both Judaic and Christian texts. However, Guides do not have wings. In fact, references of angels with wings are strictly a Christian myth, unrelated to any rational historical context.

In the Torah (the Books of Moses in the Bible), "angels" speak to Hagar in Genesis 16:9 to 16:11; to Lot in Genesis 19:1 to 19:17; to Abraham in Genesis 22:11 to 22:18; to Jacob (Jacob's Ladder dream) in Genesis 31:11 to 31:13; to Moses in Exodus 33:1 to 33:2; and to Balaam in Numbers 22:31 to 22:35.

Nowhere in any of the verses above are angels ever described as having "wings." Christian artists were apparently required by their Popes and bishops to create a myth that angels descended from a "heaven" on "wings." This myth reinforced the fable that there was a heaven from where angels descended from. It may very well be that the entire concept of a winged angel is nothing more than another of Athanasius' fables.

Indoctrination of the masses that there was a heaven was necessary to reinforce the fable that there was a consequence to a life separate from the Church's influence and authority—hell. If one refused to obey the Catholic Church, a violent death was almost inevitable.

Where Do the Guides Come From?

Guides exist in a place I call "ShadowLand" which we will discuss further, shortly. There are several possibilities as to the origin of Guides.

One possibility is that they may be beings from a far distant past who evolved beyond the desire for a physical existence millions of years ago. Intelligence need not necessarily be limited to humanoids.

However, genetic evidence leads to the conclusion we humans are the highest life-form on the planet, so far. Further, the archeological evidence indicates we humans are only a relatively recent addition to Earth's historical record. Lastly, had the Earth been home to a much older civilization and much more evolved civilization, we should have discovered some physical evidence of its existence by now—unless it was a water-borne species like porpoises, whales, or the like.

Modern science knows still knows very little of Earth's oceans. It is not a stretch to say our present knowledge of genetics is based predominately on land mammals—not ocean-dwelling life. Our knowledge of the genetic record of sea-life is miniscule. In fact, we know more about the Moon than we do about our own oceans and the life that exists there.

So, it is very conceivable that a self-aware species of aquatic life evolved way beyond our level of development long ago. Such a species could have existed, developed self-awareness, and then evolved beyond the need for a physical for—and we know nothing about it.

A second, possible scenario is that of the alien race which came to Earth in the distant past and either were non-physical beings or discovered a nexus to the non-physical realm and chose to remain there. They may have discovered that, for all practical purposes, time in ShadowLand does not exist. As long as they remained in ShadowLand, they would be virtually immortal.

There remains a third possibility. We should not rule out the possibility that the Guides were created by the Universal Mind, just as we were. With this theory, the Guides were never meant to be physical beings, but were made capable of serving as intermediaries between the physical and non-physical realms. It is true that the Guides I saw in ShadowLand appeared more humanoid than the white orbs I encountered in the hospital trauma room.

The Universal Mind may be like a super-Internet with connections available for every living being. The lower the frequency and the longer the wavelength, the more direct the connection.

During my Near-Death-Experience, I got the impression that the Guides did not want to remain in this physical realm. They definitely had a time limit on how long they could exist here. This may be explained by the wave-particle duality of a photon.

Assuming the Guides are pure photon-consciousness, they may need to remain in particle form to maintain their coherence. ShadowLand may shield the Guides' particle nature in some way that allows them to exist separate from the space-time continuum we humans inhabit.

The frequency and wavelength the Guides exist within must be in the far, extreme lower band of the electromagnetic spectrum. This conclusion is based upon my meditation experience. Whenever I meditate, I feel my consciousness rise and my awareness speed up as my body functions relax and slow down. I also feel a slowing down of my body rhythms, which must be caused by the lengthening of my brain waves.

Higher consciousness must be indirectly proportional to the length of the brain waves. This makes sense when you consider that your active mind is driven by sensory inputs that rev-up your awareness. This increased sensory awareness, necessary for survival in the physical realm, nonetheless is the exact opposite of what is needed to connect with the

Universal Mind. What is needed to connect with the Universal Mind is stillness of thought and clarity of self.

Humans' limited enlightenment may not allow us to possess the needed frequency for more than limited awareness of the connection. The Guides may have evolved or become enlightened to the point where they possess the necessary frequency to simply tap into this network at will. This would also allow them to connect with us at will, whenever they deem necessary.

What and Where is ShadowLand?

First, how do I know? I was there in ShadowLand between this life and my last. Meditation and my "Near Death Experience" allowed me to remember what happened to me while I was there. This and my other books on the Universal Mind and Reincarnation are dedicated to sharing with you what I experienced.

The Guides interact with us directly as they process our unique individualities ("souls") from one life to the next. They communicate with us telepathically, while we are in ShadowLand as well as here on Earth.

Therefore, they must exist within the same universe as we do. Therefore, ShadowLand must also exist within this universe. Therefore, the laws of physics must apply in ShadowLand as they do everywhere else in this universe.

As an example, I observed ShadowLand to have physical-like structures. There was an altar-like stage with a mantle upon which sat several ornate chalices. We will discuss the nature and purpose of these chalices shortly.

Several large mechanical devices were situated behind the altar, as well. They appeared involved, somehow, in transporting us back to the physical dimension and inserting our consciousness in the fetus of our future bodies.

However, this does not mean ShadowLand possesses any form of physicality. The altar, the machine-like structures, and even the appearance of the Guides themselves may have been merely mental projections created in our minds by the Guides for the benefit of our underdeveloped consciousnesses.

From these observations, we can make a few assumptions, however. As religion is no longer relevant to our analysis, ShadowLand should not be confused with any mythological heaven. Therefore, its existence must be that of a cosmological construct—a structure built by the Universal Mind, just as we were.

As we have established that consciousness is a creation of photon entanglement and nonlocality that exists throughout the universe, we know that the size of ShadowLand is at the quantum level. Therefore, ShadowLand may be no larger than subatomic particles.

It need not be. The Guides in ShadowLand process us while we are totally incorporeal—nonphysical. When we are in ShadowLand, existing in between our physical lives, we are pure thought energy. Therefore, ShadowLand need not be larger than a micron or smaller.

Because of its dual nature as both a wave form and a particle, measuring the diameter of a photon is next to impossible. We measure its wave form in terms of frequency and wavelength. However, when photons are processed through ShadowLand they lose their wave-particle duality and exist as particles only. While photons are in ShadowLand, they no longer move as they would in "normal" space/time.

To provide some perspective, 70 microns is the diameter of a human hair. The naked eye cannot see anything smaller than 40 microns. A white blood cell is 25 microns in size. Coccus bacteria is only 2 microns in size.

Source: Jennifer McAlpine, "How Big is a Micron," Benchmark, at bench.com, 2020.

It is also significant that the Guides process physical beings through ShadowLand very quickly. So, there must be something inherent in the physical dimensions and the Guides which prevent us from safely remaining in ShadowLand for too long.

It may be something as simple as the Guides' non-physical nature is the result of existing at a wavelength and frequency that we cannot—and remain able to return to the physical realm.

The only explanation for this can be that ShadowLand is traveling at or close to the speed of light. If true, this means that most of the beings in ShadowLand (Guides and us) are reduced to photons. This implies, I believe, that when we exist in ShadowLand, we are almost pure thought. Thought is consciousness in action. Thoughts are things. Thoughts have momentum, they have energy.

When I died in past lives, the Guides always instructed me to "think music." Sound or vibration in some frequency and wavelength must have an effect upon our successful processing in ShadowLand and rebirth back into the physical realm.

Thinking of music may serve more than one function. First, it may be meant to calm us as we pass through what must be a traumatic experience for most of us. Second, thinking of music while there may serve to center our consciousness, like during meditation. This centering may be necessary to properly adjust our past Karma with our future lives.

It is possible that much of what the Guides do is creating life-gravity-centers for our next lives. Life- gravity-centers are events in our lives that we are fated to experience. These events cannot be avoided. Regardless of our life choices, we are "drawn" toward specific places and moments of time to experience certain events.

These life-gravity-centers pull us towards them like black holes draw in surrounding stars and planets. They exist to bring us to those who would be our future mates, children, loved ones. Life-gravity-centers are also Karmic in origin. Life-gravity-centers also are created in future lives to resolve past-life Karma.

Why Don't We Remember ShadowLand?

Memories of our time in ShadowLand, along with memories passed down from our past lives, are implanted in the spinal cord of the fetus of the person we are to be reborn as. These past-life (episodic) memories migrate up the spinal cord into our brain as our fetus develops. By the time we are born, these past-life memories are already encoded by our limbic system and stored deep in our brain as an imbedded neural-memory.

By meditation I have been able to remember my past-lives. Certainly, my "Near Death Experience" had an effect. But I believe whatever effect my NDE had was limited to being the motivation for me to meditate to begin with. Meditation is the key to past-life memory, not nearly dying.

If I can remember my time there, anyone can. I don't think I enjoy any special privilege. Remembering events is dependent upon how much you are willing to do so.

"If you seek, so you shall find."

Those who have not remembered their past experiences in ShadowLand, just as with their past-lives, really do not wish to remember or do not believe in it. Both fear and disbelief are the only "door-closers" preventing one from remembering. I do not believe the Guides have anything to do with it.

How Do We Determine Our Rebirth?

How we are reborn is a cooperative process between us and our Guides. We do not get to choose what family we are reborn into. But we are permitted to choose between certain options.

For example, the Guides may offer you one of several "arranged" or "preordained" lives. This process is couched in ritual. At this point, we humans are just the photonic-encased memories of our past lives. Shall we refer to this remainder or residue of our past lives our—soul? Let's do that for simplicity sake.

Each soul is escorted by two Guides to the ShadowLand altar we discussed earlier. The soul is instructed by its Guides how the ritual is to proceed. Resting on a mantle upon the altar are several ornate chalices. Each chalice represents a "preordained" fate the Guides have created for that particular soul's next life. The soul is given the opportunity to choose between the offered chalices as its next life.

On the exterior of each chalice is a carved human face. Each "face" represents the actual face the soul will assume should that chalice be chosen. This face is an actual representation of the face of the person who would become if that particular chalice were to be chosen.

One may be offered as many as seven such chalices. The proffered chalices may include different races, genders, and nationalities.

The soul is instructed to take a sip from each chalice. Sipping from a chalice gives one a brief vision of the life symbolized by that particular chalice. These visions are not an overview of each possible life. Instead, they are isolated, random moments in time which will be experienced during that particular life.

This is risky business. Imagine a vision of being Gen. George Patton before he was paralyzed from the jeep roll-over after WWII; of being John F. Kennedy before Dallas; of being Amelia Earhart BEFORE that last plane ride; of being Princess Diana before that last night in Paris; of being John Lennon before he signed his last autograph; or of being Yeshua before his alleged crucifixion. What one is shown by the chalices is not necessarily the whole story of one's future life.

These are fully preordained lives; created by the Guides for their own purposes. One has very little free will with such a life. From birth to death, such lives have "fate" written all over them.

Some may know from early childhood what kind of career they will pursue. Others may not realize their life's fate until it "hits them between the eyes" in their teens or early adult years. Usually, this sudden realization of one's fate results from a single or inter-related series of events which were "preordained" by the Guides and embedded into our lives by those machine-like structures located behind the ShadowLand altar.

These machines create fate. The embed deep within our consciousness an irresistible impulse to do something at a specific moment in the future. Our actions then create a behavioral gravity-well like a black hole. Causality becomes its own genesis.

People born of these chalices are possessed with an irresistible inner drive to achieve their "mission" in life. Their free will submits to an irresistible inner impulse to manifest in physical reality the dream embedded within their consciousness by the Guides—and their own choice of chalice.

Caution is the keyword here. The Guides can create our fates for us. How they do it is not nearly as important a question as WHY they do it.

What is clear, however, is that they can and do control our destiny. To ignore the reality that they must also have their own secret agenda for us is not merely ignorant, it is unworthy of us. What is their agenda? We will discuss that shortly.

But first, we must understand that not all souls are offered the choice of chalices. Those not offered a ShadowLand chalice fall into two groups: those who are given a choice o how they will be reborn and those who are not. While I do not know for certain wha criteria determines which group one falls into, my assumption is that it somehow depends upon the individual's level of self-awareness.

Think about it. Would a responsible parent allow a child to choose on their own whethe to go outside on a cold day with or without warm clothes? Would any responsible teache allow his or her students to decide on their own whether to come to class or not?

Of course, not.

Shouldn't the freedom to choose a particular course of action in a particular matter be reserved only for those who have proven themselves worthy of making an intelligent and reasoned choice? Isn't the definition of "responsibility" that of being "able to respond" appropriately to the situation at hand?

That may be how the Guides determine those who merit the freedom of choice in preferences for their next lives. For those who do merit the freedom of choice over some aspects of their next lives, the process is very simple. It is as basic as filling out a personal-preference form for an online dating site.

Each soul is offered an opportunity to provide their preferences, such as race, gender, nationality, the geographic region one wants to be born in, type of parental influence, one's career-interests, etc. The Guides then try to match the soul to a family which most closely fits its chosen preferences.

We do no choose which family we are reborn into. That is processed by the Guides. Furthermore, we are stuck with the biological and genetic traits we are given by our "birth family" the Guides place us in. Depending upon how many preferences a soul has, the Guides may not be able to find a "birth-family" that matches every preference. In that case, the Guides find as close a match as possible and there you go.

For example, let's say we announce a preference to be born in the Midwest USA. This is a region settled predominately by central and eastern Europeans. The heritage of this region is impacted by certain genetic traits, including susceptibility to certain diseases. One whose preference is to be reborn in a particular region will have to accept the genetically-controlled physical features and illnesses prevalent within that regional demographic.

We can also be born as any gender. Our Guides leave it to us as to whether we prefer to be born as one gender rather than another.

Yes, we can be born as either gender as well as any race. We can also be born of any nationality. However, the more one is reborn as one gender or race or nationality, the more likely one will choose to be that in future rebirths. I have no preference and have no problem being reborn in my next life as a woman. I believe it would be fun, though my wife says otherwise.

However, due to biology, one's choice of gender will affect which human family one will be born into. In other words, one cannot be born into a particular family as a male if the male parent of that family only produces female sperm. Both we and our Guides must still accept the rules of universal laws, including genetics.

What Do the Guides Want?

We discussed earlier how the Guides can create fates for us. They can also influence our reactions to events without our knowledge. These facts lead to the inevitable question of "why?" What is their motive? What do they want from us?

Before you begin running around in circles screaming the world is about to end, get your head out of the sand. If the Guides wanted to destroy us all they needed to do was leave us alone. Without their guidance and left to our own biological and psychological impulses, we would have self-destructed long ago.

Whatever they are and whatever their origin, our Guides have but two apparent motivations: to protect us from ourselves as a species and to aid in the development of our personal consciousnesses.

CHAPTER 23

Past Lives and Future Hope

Now it is time I shared with you a my most recent past-life. Memories of this surfaced while I was finishing my graduate work at Kent State University in 1979. At the time I have the dream, I am in trying to choose whether to go to law school or seminary.

Despite my better judgment, I was strongly drawn to becoming an Episcopal priest. Though every ounce of my intellect told me religion is a fraud, something deep inside drew me towards a life of faith and contemplation in the service of humanity as a priest.

It is then, as I struggle to choose a direction for my life, I experienced the same "waking-dream" every night for two weeks. This dream haunted me. It was more than just a dream. It was also a prophesy…which came true many years later.

The dream was so vividly real I woke up each night sweating profusely; physically exhausted and emotionally spent. At the time, I was involved in a very serious relationship. My nightly horror-dreams affected us both. But neither of us had any choice. It became a macabre routine. Every day, my girlfriend would ask if I was going to have the same "war-dream" again. I would tell her there was no way I could have another such dream.

But I would be wrong. Every night, the same dream would play out its horror. I would go to sleep expecting it to happen again. The year of the dream (1979) is very important, as you shall realize shortly.

The following is the "waking-dream"…

****

It is WWII and I am a soldier. Was I a Marine? Not sure. I was a coronet player before the war, playing cool jazz in clubs. But there is no sweet jazz now. Not now. Not here.

Here is hell. Me and my buddies are being chewed up on a beach on some island the "brass" assured us would be a cakewalk to assault.

Some cakewalk. The beach is a slaughterhouse and we are its meat. The beach has disappeared. The sand is buried beneath our bleeding flesh. Dead bodies cover the beach like a blanket of broken flesh; layer upon layer.

Our enemy is dug in deep inland, in the jungle, and hitting us with heavy mortars and machine guns. The mortars make a distinctive two-sound…one sound when it shoots, the next sound when you die.

Thump….BOOM! Thump…BOOM! Thump…BOOM!

Each mortar shell blast blows my friends—or what is left of their bodies—high into the air in a macabre tapestry of blasted sand, flesh, blood and bone. With a twisted fascination, I find myself staring at the bodies cartwheeling through the air like acrobats in a death-circus.

War as art form. Definitely Dali or Magritte. We cannot hide. If we stay, the mortars get us. If we get up and try to run for cover, the withering fire of machine guns knock us down like cord wood. There is but one way to survive—crawl under the dead and dying.

I crawl. I burrow beneath the torn flesh and bone of my dead and dying comrades.

A mortar must have hit me. I did not hear it. The explosion must have torn my ear drums. I wince seeing the blood, but then realize it's not mine. By dumb luck, I have survived. Why am I not dead like my friends?

Evolution does not exist in war. War is chaos. Life is no longer "survival of the fittest" but survival of the luckiest. Can "luck" be in my DNA?

Life is lived from mortar round to mortar round. If you live long enough, you learn to tell where a mortar round will hit by its sound as it falls through the air. If the sound is a high whine, the round will overshoot—you live. If the round is a low-sounding groan, it will undershoot—you live.

But if the mortar round is in the middle-range, in between the whine and the groan, it is headed for you. Generally, each round took about six seconds from the "thump" of it being shot and the inevitable blast. You say your prayer and wait. Some call it the "doppler-effect." That day, on that beach, we called it "the six-seconds."

How we secured that beach, I do not consciously recall. I think I compartmentalized it in my "monster closet." Once the beach is secured, time drags. We're waiting for resupply before we push inland. The Seabees are repairing the airfield. It's sweltering-hot and so humid we wear only our battle pants, boots and socks. Shirts are only used as towels to wipe away the sweat.

This leaves us prey to the ever-present, bloodthirsty sand fleas. I don't know what is worse, the constant sweating or being constantly bit by sandfleas.

Personally, I hated the sand fleas more. I eat every sand flea I catch. It's only fair. They bite me, I eat them. I can taste their blood. I love the taste of sand flea blood. All of us are living on the edge; the edge of life, the edge of sanity.

Bored to death, I take my weapon and trek into the jungle, alone. Maybe I can pick up a trail leading to the enemy camp. None of us look forward to their late-night assaults. I don't plan on going far; just a few hundred yards in. They're in there somewhere....hiding....waiting. Maybe if I can find their camp, we can surprise them.

I tuck my submachine gun into the crook of my right arm and walk in; silently stalking. About a hundred yards in, I find what I'm looking for. I am walking on a narrow piece of worn earth, no more than two feet wide. There are no leaves covering it. It's fresh, real fresh.

I go stealth-mode-quiet. I follow the trail as it turns a corner and then curves around a huge banyan tree. As I reach the other side of the banyan tree, I see him.

A Japanese soldier lies on the ground, leaning up against a tree. He does not notice me. He's busily writing in a journal. His rifle lies on the ground beside him. He's not even wearing a helmet. For all anyone could tell, he could be enjoying an afternoon picnic; not taking a break after the morning slaughter.

My finger automatically finds the trigger. I have the "drop" on him. But he keeps writing in his journal, oblivious to my presence. He appears so composed, so...serene.

This is intolerable. His composure demeans the deaths of all my buds whose bodies lay in bloody pieces on the beach. He needs to know I am here. He needs to see my weapon at the ready. He needs to know I have the power of life and death over his sorry ass. He needs to know he will die at my hand. He needs to see it. He needs to feel it.

I need to see it. I need to feel it.

I cough.

His head rears up like an animal realizing it is prey to predator. Our eyes meet. He's wearing round spectacles, in round, black frames. The soft, round face. This is no soldier. More an academic than warrior, I think. We could have been friends, before.

Our eyes never leave contact with each other. His left hand makes a small, furtive move toward his rifle. I shake my head. This is not how I want it to end. He does not have to die. Or, does he?

"Touch it! Touch it!" My mind screams silently.

His hand touches the rifle. My finger pulls the trigger.

As my bullet speeds toward my enemy's heart, so too does my soul. My soul lives on the tip of the bullet, whistling its way through the humid jungle air. Something is flying from out of his chest. Something from him flies towards me as I fly towards him.

Our souls collide. They pass through each other like two opposing winds. His wind carries me away with it. Unwanted, his life flashes before my mind's eye. Unwanted, his hopes, his dreams, his loves, fill my mind.

My reaction is very simple, very pure. I go insane.

Why?!

Removing all my clothes, I run naked back to the beach, screaming all the way.

Why?!

My buddies try their best to talk me back to some sort of sanity, but I am too far gone. They send me back to a troop ship and lock me in a room. The next thing I recall, I am in San Francisco. There is a huge naval base there. The hospital I am in triages all the wounded arriving from the Pacific.

I am sitting on top of an ice-cold, metal table, waiting hours to be seen by a doc. Then, I spend more time listening to this moron who has no clue what to do with me.

I want to kill him. I have no weapon, just my own hands. That's all I need, though. Then I decide not to. So, I just hurt him, a lot. Guards came in to stop me. They have not been where I have been. They have never done what I had to do. I hurt them, too, a lot.

I don't recall much of San Francisco after that. Sometime afterwards, I am sent for long-term rehab at a mental clinic. The clinic is next to a large VA hospital in a quiet, backwater place in the sunny, warm South. The mental clinic has two floors. The first floor is a canteen (a small cafeteria and gift shop) that looks like a Swiss chalet. It's ceiling even has painted, wooden rafters—quaintly beautiful.

I am walked around the grounds daily by a very pretty, young nurse with red hair. The path leads to a pristine but small bay. It is a very soothing place to live. I am told to seek soothing. Soothing is good. Hurting people is bad. I have a monster hiding in me and it mustn't come out.

So, they say.

Months pass and the war is over. The monster inside me is asleep. It has been asleep for a long time now. The docs say I am well enough to be the clinic's overnight janitor. I am very proud of myself. I have made wonderful progress. I am one step closer to leaving.

My first night as janitor, I go down the elevator with a pail of cleaner and a mop to clean the first floor where the canteen is. It is my first night on my own. I am very proud and confidant that I am close to being released. The elevator takes me downstairs to the canteen. The elevator door opens. I take my mop and bucket and walk out onto the canteen floor. I turn on the light.

NO!!!

There, on the floor, are the tiles of fascist evil! Swastikas everywhere on the floor! How can there be swastika tiles on the floor of an American VA clinic?! My brain tries to understand what is too horrific to imagine.

The implications were obvious. If the VA had swastikas in their floor tiles, it had to be a message that all the American government AND its military were being commanded by Nazis!

Obviously there was a conspiracy involving everyone in national authority. But fighting such a large conspiracy was hopeless. No one would believe a mentally warped veteran such as myself.

So, I did the next best thing. I rushed to the janitor's storage closet. It was locked, so I broke the door to get inside. I filled the bucket with lye and every other chemical I could find in the janitor's storage closet. Then I poured it over the floor by the elevator door.

The next day I am released from the psych ward. I am told I have been "healed" but I know the truth. They didn't want to keep me anymore. I had become uncontrollable.

****

That is the dream I had in 1979.

Time passed. I graduated law school in 1982, passed the Florida Bar exam in 1983 and moved from Ohio to Florida to begin my life as a lawyer.

In 1991, I began my twenty-five-year career as an Assistant Public Defender. My first assignment as an Assistant Public Defender, in 1991, was to represent individuals who were facing involuntary detention under Florida's "Baker Act" statute.

In Florida, a person can be detained against their will if they are found by a court to be a danger to themselves or others due to a mental illness. Formal evidentiary hearings are conducted, in front of a judge, with testimony from both mental health professionals as well as the patient concerning whether the patient requires involuntary detention for treatment and for how long.

Pinellas County, Florida has several such hospitals, and "Baker Act" hearings are conducted weekly at each of them. One of these mental health hospitals was a two-story building located on the campus of the Bay Pines VA Hospital. The clinic was located beside picturesque Boca Ciega Bay, on the Gulf of Mexico. It was locally referred to as "the canteen."

As soon as I entered the building, I felt like I had stepped back into my dream from twelve years before. The entire time I was there, I felt as though I was only partially in my body.

Sure enough, the first-floor canteen was a quaint, little cafeteria and store. Wooden beams crisscrossed the ceiling like a Swiss chalet. It was even more quaint than it appeared in my dream.

The mental health clinic was still on the second floor. On the floor, in front of the elevator, were hundreds of tiles. Each tile was imprinted with the same broken-cross symbol the Nazis stole as their symbol of fascist and racist hate—the swastika. The stain I had poured over the tiles was still there, as well.

Telling the clinic staff that I knew how the floor got stained did cross my mind. It did seem the only appropriate thing to do.

But I didn't. Too much time had passed—almost fifty years. The hospital staff would have had no idea what I was talking about. They might have "Baker Acted" me themselves.

What I had thought, in the 1940s, to be a Nazi conspiracy was actually an historical coincidence. When the clinic building was originally built in the 1920s, following WWI, the swastika design was still considered an auspicious design used in many buildings.

In fact, the design was favored by many ancient religions and cultures around the world. It had been used as a symbol of divinity and spirituality amongst peoples of India, Nepal, China, and Japan. When built in the 1920s, the "canteen" had no relationship at all with Nazism or fascism. My conspiracy theory was the raving of a mad-man—the monster inside me.

In 1979, how could I dream about a psychiatric hospital—and its swastika floor tiles—when I had no way of knowing (in this life) that such a hospital existed? Even more astonishing…imagine the odds of this my dreaming in 1979 of a hospital that I'd never been to before (in this life) and then actually working there as an attorney representing clients in 1991?

Speaking of odds; what are the odds that I dream in 1979 about a mental health clinic I was treated at in the years following the Second World War with such detail? How did I end up twelve years later, in that very same clinic, defending mental health patients—like I had been, myself, so very long before?

How does that happen? Coincidence? You must be kidding me. Coincidences like that do not exist.

Allow me to explain how…

The dream had been stored in a neural-synapse, deep in my brain, as a past-life memory. What triggered the memory to come to the surface? I believe during the time we spend in ShadowLand, that the Guides program our past-life memories to come to the surface of our consciousness at certain times in our next lives. These memories apparently are necessary to our spiritual growth.

The ability to have "past-life" dreams is dependent upon several factors. Please keep in mind that I can only give you what I have experienced myself. Though it worked for me, it may or may not work for you.

First, I wanted to remember my past lives. I was highly motivated. My NDE gave me a very strong reason to meditate. I wanted very much to remember my past lives.

Most say they want to remember but are too fearful of what they will re-experience by reliving those memories. The mirror of your mind does not lie and cannot be fooled. You can meditate for the rest of your life trying to experience your past lives. However, any inner doubts you have about re-experiencing past lives will block you from remembering any of them.

Continued, proper meditation is the key to remembering past lives. Our consciousness is like the soil, the earth beneath the feet of our sensory-absorbed lives. To reach the Universal Mind, we must meditate to "dig" down through the layers of consciousness, to bore a opening between the Universal Mind and our conscious awareness.

We possess three layers of consciousness. The most superficial layer of consciousness is our perceptual awareness. This is what is called sensory perception. Our minds are consumed by these images every day. They are transitory, fluff, flotsam, bubblegum for the senses.

Symbolically, this level of our consciousness is equated with the organic (humus) and topsoil layers of the earth. These layers are easily disturbed and uprooted. Yet they are full of life. Their seeds are scattered everywhere, but their roots are short.

The second layer of consciousness is reflection-awareness. Our minds reflect upon what we have experienced through our senses. This reflection gives our minds the opportunity (through the limbic system) to choose whether to apply an emotion to the experience and to what intensity the emotion should be.

Those events which are not given an emotional loading or coating, are forgotten almost immediately. Those experiences which do have an emotional load given them, are processed to a synapse and neuron for storage. These become our memories. Our strongest memories create our identity—our individuality.

Symbolically, this level of consciousness is like the eluviation and subsoil layers of the earth. This layer contains the remains of plant and animal life from the recent past. They represent our memories of the events, experiences, and emotions we have experienced in this life.

The third layer of consciousness is very deep and connects us with the Universal Mind. It begins with the end of our present life experiences. It then travels through all the memories which have been preserved from our past lives. Then it proceeds deeper still until it reaches the Universal Mind itself.

Symbolically, this third level of consciousness represents the bedrock covering the heated mantle and inner core of the Earth. The deeper one "digs" down into consciousness through meditation, the more information of one's past lives comes to the surface.

However, if subconsciously, one is reluctant to experience past life memories, meditation becomes a waste of time. For every "shovel-full of soil" one digs through by meditation, the mind's fear of discovery fills back up. This is why it is pointless to force meditation. When the time is right, meditation is an effortless experience and joyous journey.

There is nothing preventing us from visualizing our past-life memories—but our own fear of what we will experience. From personal experience, I can tell you, they can be profoundly enlightening—and traumatizing.

Some past-life memories come to our awareness in waking-dreams, like my war-dream. Others come in a strobe-light flash during meditation.

How To Live with Past-Life Memories

My dreams of my past-life as a soldier fighting in WWII in the Pacific were very traumatic. Without the proper preparation, I could have been psychologically damaged, even become suicidal. Seeking past life memories is not recommended for anyone unless they have been properly prepared ahead of time with years of proper meditation and a stable psychological foundation, which most people here in the West lack.

Past-life dreams may appear following years of meditative preparation, or they may simply begin spontaneously without warning. In either case, it is not recommended to pursue past-life dreams unless under the supervision of a trained team; including a meditation expert or spiritual teacher and either a licensed psychotherapist, psychologist or psychiatrist.

We humans enjoy many gifts from life; including physical, intellectual, and artistic abilities wonderous to behold. But not everyone is talented in sports, art, or music. Each of us is born with both strengths as well as weaknesses. Some can throw a baseball over 90mph. Most cannot. Some can remember past-lives. Most do not.

Now, let's assume you have actually begun to experience past-life dreams. You have your "team" of meditation and psychological counselors on-board and you are ready to proceed. You are probably wondering how to deal with the real-world complications your past-life dreams may create for you. You likely have many unanswered questions.

Experiencing past-life dreams can create problems. Imagine having past-life dreams if one is Christian or has Christians in the family who do not believe it past-lives. You may wonder whether you must do anything in this life to resolve any past-life Karma; whether you can control how often you have them; or whether you can ever confide to anyone about them.

Most humans do not have past-life dreams. Come to think of it, I cannot recall ever meeting anyone who admitted to me they had even one past-life dream themselves.

But those of us who have had past-life dreams mostly keep that fact to ourselves. We experience them in private. We cope with the experience of each past-life dream as best we can, on our own. That can be difficult, as many of our past life-dreams involve some form of emotionally traumatic event.

Having past-life dreams is like living two lives at the same time. Our first life is the one we live out day-to-day in our present lives. But then we also must deal with our past-life memories, some of which give us as much emotional trauma as would any event in our present lives. So, I have compassion for anyone who has had to deal with past-life dreams.

Having past-life dreams is not for the weak of heart—or mind. This fact has made it easier for me to accept the reality that we "Past-Life-Dreamers" are a very rare minority.

It did not take me long to learn it is pointless to share past-life dreams with others. Trying to explain to others about experiencing a past-life dream is akin to discussing a color photo to the blind. There is no shared frame of reference.

Sharing one's past-life dreams with family or friends (even close friends) risks being thought of as weird. Let's face it, most American's are blind to past-life reality. Trying to explain past-life experiences to most people is like trying to explain color to a blind person.

From time to time, I have opened up to past lovers and friends about my past-life dreams. Almost always, it does not end well. Most change the subject immediately. I must admit, their perception of me changed as a result. They had great difficulty dealing with it.

This applies even to loved ones. My wife and I have been married for almost thirty years. She is my best friend. We share everything. And yet, every time I discuss my past-lives with her, she looks at me like I'm from somewhere north of Mars and starts giggling. But she has helped me understand why this is so. It is very difficult for those who have never experienced a past-life dream to wrap their brain around the concept.

Should you tell your co-workers about your past-life dreams? I wouldn't recommend it. Even as open-minded and educated as professionals claim to be, I don't recommend sharing such experiences with doctors, lawyers or the like. Just because I sometimes opened up about my past-life dreams with other lawyers I worked with doesn't mean you should. You risk being ostracized or worse.

Most Americans, and other Westerners, are indoctrinated from birth that past lives are contrary to Christian theology. They are raised to believe what the Church tells them.

People brainwashed by a life lived under Christian lies feel uncomfortable when shown an alternative reality. It rocks their world-view.

The process of remembering past-lives is inextricably intertwined with an evolving relationship with the Universal Mind by an expansion of your consciousness. If you have had past-life dreams, your consciousness has already reached a level where you may have already had contact with Guides independent of the dreams.

You have three sources from which to seek personal advice. First, you have your own past-life experience to draw from. Second, ask your Guides to help you. Last, go straight to the source—ask the Universal Consciousness itself. It may reveal to you insights you had not considered.

But do not let your past-life memories control your present life. Just as a beaded necklace is made of separate beads connected on a string. Each bead of a necklace is as important as any other. Each should be appreciated as a separate entity within the shape of the necklace.

So, too, must be each of our separate lives. Each life is unique, full of its own unique challenges and rewards. You cheat yourself out of the joy of life by limiting your present experience by over-indulgence with your past. Reincarnation is meant to free our consciousness from the negativity of our past lives and give us new hope and opportunity for spiritual growth with each new life.

If you have past-life dreams, consider it as a very special gift. Your consciousness is progressing as it should and has evolved to a greater realization of the universe—and your place within it.

Good Versus Evil

Fate Versus Free-Will

Yes or no; hot or cold; young or old? Love or hate; yin or yang; in or out; up or down; winners or losers; rich or poor; right or wrong; good or evil? Our opinions reflect the duality of our physical existence.

Our physical existence—and the physical senses with which we experience that existence—tell us life is an either/or experience. This is evolution. It is in our DNA as surely as the words “SAFE” and “NOT SAFE” are branded upon our past-life memories since the beginning of life on Earth.

But like our five senses, our memories lie to us. Physical existence may be a plus-minus duality but consciousness is not. A Ukrainian fighting for his homeland comes across a young man from Russian city of Nizhny Novgorod Oblast who has been ordered to advance on the “enemy” or he will be shot as a traitor to the State.

Who is right? Who is wrong?

Which man is good? Which man is evil?

Was the young soldier I happened to discover, leaning under a tree during WWII, good or evil? Was my shooting him right or wrong?

The Each of us live out our lives in a prison of subjective experience. Because we cannot feel the thoughts and feelings of others, we are left to believe the only thoughts and feelings that matter are our own.

We are separate from each other physically. As a result, we are separate from each other emotionally as well.

For almost two thousand years, the UnChristian Church has taught that life is an either-or proposition. For almost one hundred generations, humans have been indoctrinated that life is built upon the foundation of good versus evil and that we are born sinners condemned to a life of evil unless we are "saved" by the Christian god and his servant-son Jesus.

When we have the courage to look into the mirror of our own souls, what we find looking back at us is the face of…not god or satan…but ourselves. The face staring back at us is the present image of who we are—created by the memories of hundreds of past-lives we lived in this physical existence.

But even that face is an illusion. Again, to the extent that we focus our attention upon our past lives we fall into the trap of separateness. The greatest challenge of life is to break through this prison of separateness and experience commonality--oneness. Meditation is the answer. The only way to develop an increasing realization of oneness is to meditate.

Oneness is the Universal Mind. The Universal Mind does not teach either-or life scenarios. The Universal Mind does not teach. The Universal Mind is effortless being. The Universal Mind does not need to "try" to be anything. It simply is.

When you realize you are what you are, you need not make any effort to be that which you are naturally. Again, that is the purpose of meditation—effortless realization of our true nature. Oneness is not taught—it is experienced.

The Universal Mind does not experience good and evil. Both are illusions we created out of our psychological dependence upon a physical existence based upon duality.

Does this mean we are not responsible for our choices? Does this mean we cannot be held accountable for our actions? The answer to both is "no."

While our genetic heritage and birth environment predetermine much of our life, we still have free will to make choices. Our choices create causality loops we call Karma. We can become trapped in these causality loops, vicious cycles of emotional chaos which are self-destructive. Enlightenment releases us from these causality loops, allowing us to evolve—to spiral upwards in conscious awareness—and ever more productive lives.

Do We Have Souls?

Yes, we have a "soul." But it is not what Christianity teaches us. When I went through Catechism at age twelve, I was taught that my "soul" was created for this life and this life only. The only way to achieve eternal "paradise" was to believe the scriptures, believe the Nicene Creed and beg the Lord for forgiveness of my "sins." Only then would my "soul" qualify for immortality in "heaven." Otherwise, my "soul" would end up in "hell," forever.

See how easily the dogma becomes indoctrination? The reality is much different. The term "soul" actually refers to the photon-shell of memories we take with us from life to life. These memories are the multi-life experiences that make us who we are. They are our unique individuality—our soul. Your soul has been tied to you, and you to it, since the beginning of the universe. Your soul will exist until the universe collapses to form the next big bang—and maybe even longer as each universe blends in to the next.

Your soul is your responsibility. It is up to you to feed your soul with the "bread" and "water" of enlightenment. Just as your body suffers from bad food, your soul will suffer from poor life-style and negative-hateful thinking.

None of us are perfect. We all make mistakes. But the instinct for learning, growth and renewal is what makes us human. Never give up, never lose faith in who and what you are—a child of the universe.

Is Time Travel Possible?

"Never say never," is one phrase. "Anything is possible," is another. Someday in the future, humanity may be able to bend natural law to allow time travel, but until then, the answer is "NO."

The reason for this includes two consequences of the Second Law of Information Thermodynamics: the "nondeleting" and "noncloning" principles. The father-son team of Rysand and Michal Horodecki are just two of an entire family of physicists from Poland's University of Gdansk. Together, they reformulated the Second Law of Thermodynamics to conform to quantum mechanical principles. Their redefinition of the Second Law of Thermodynamics is as follows:

> "In quantum mechanics, one cannot either delete or clone (quantum information). The Second Law holds: entropy cannot be decreased in a closed system. Again, the Second Law can be stated in a stronger way: that eigenvalues of density matrix do not change under quantum evolution. And this is actually the information quantum principle."

Source: Horodecki, H. and Horodecki, M., et al, "No-deleting and no-cloning principles as consequences of conservation of quantum information." arXiv, 2003.

Just so you know, eigenvalues are merely mathematical components used in linear equations commonly found in quantum mechanical computations. All the Horodeckis are trying to say is that quantum mechanical principles apply to all changes in any system which obeys thermodynamic laws:

"In a closed system, one cannot destroy quantum information. Quantum information can only be moved from one sub-space location to another. In a closed system, even orthogonal (opposite) states cannot be deleted alone, given unitary evolution. Once entropy or information has increased, it cannot be decreased—it can only be transferred to a separate location. Quantum information cannot be "leaked out" while keeping original information intact…Once information is increased, it cannot return to its original state."

Source: Horodecki, above.

Now, let's apply these principles to the question of time travel. Let's say you want to go back in time to the year 1922 and kill Hitler before he can rise to power. Of course, that is before you were born.

Traveling back in time would result in the deletion of all the information you learned in your present life. The Second Law's "no-delete" principle prevents you from losing that information. Even if your body was successful in arriving back in 1922, you would be brain dead.

All the information you had gained during this life would be no longer inside your brain—it would all be somewhere else. Going back in time requires one to lose the information gained in their life in the years after.

Going forward in time presents the same dilemma. Even if you were able to go forward in time, you would never be able to return. The Second Law would prevent you from going back to your starting time with more information than you had when you began.

Time travel requires not only the ability to transport physical bodies through quantum space-time, but consciousness as well. Consciousness is self-awareness. Self-awareness creates its own information-state. Information, as a state of existence within a closed system, must obey natural law.

Can We Be Reborn as Different Races, Genders, and Nationalities?

Yes, we can. I have experienced past-life-dreams of being Roman legionnaire, an English peasant who fought for the House of York against the House of Lancaster in England. Other past-life dreams include lives as an English mercenary, a Native American, as a coronet player who fought in the Pacific during WWII and at least one past-life as an African-American.

Human sexual preference is a function of our DNA.

"In theory, humans and other animals who are exclusively attracted to others of the same sex should be unlikely to produce many biological children, so any genes that predispose people to homosexuality

would rarely be passed on to future generations. Yet, same-sex attraction is widespread in humans, and research suggests that it is partly genetic (and environmental)."

Source: S. Reardon, "Genetic patterns offer clues to evolution of homosexuality," *Nature* at nature.com., August, 23 2021; citing B.P. Zietsch, et al, *Nature Hum. Behav*. (2021).

The Universal Mind does not judge anyone by sexual preference. Nor does the Universal Mind judge anyone by their religion, heritage, gender, political ideology nationality, or creed. The Universal Mind is an objective source of life and consciousness It does not hate or love in the human sense. Rather, the Universal Mind is like that of the *Star Trek* character, Spock, and other Vulcans; totally objective and unbound by human emotions.

Instead of emotions, the Universal Mind operates on the natural laws created during the earliest days of our universe. This includes Karma, which is only the social derivative of Newton's Law of reciprocal motion in a vacuum (every action creates an equal and opposite reaction).

Therefore, anyone can be born of any race, religion, political ideology, creed, gender, national origin, or sexual preference—without fear of any judgment whatsoever. Ironically, the only judgment the Universal Mind allows regarding human rights applies to those whose hate creates bad Karma which will inevitably come back to haunt them in this and future lives.

The over-arching principle is that what we do with our own minds and bodies is our own business and no one else's—unless our behavior harms another. Harmful behavior to others creates negative Karma which sticks to our consciousness like sand inside a shoe. We can plead "not guilty" in court as much as we like, but the sand remains stuck in the shoe. We know what we did; and we know it was wrong.

This includes sexual preference. The LGBTQIA community still suffers discrimination, physical violence, and emotional abuse. Ironically, this discrimination, physical violence and emotional abuse is often by those whose ignorance and hatred toward anyone different than they are cloaked in the hypocritical garment of "religion." The Universal Mind does not judge anyone by sexual preference. Such oppression is wrong.

In a perfect world, each would enjoy the freedom to experience life as they are inspired to do so, from within their own consciousness and their personal relationship with the Universal Mind. Alas, such a world is still a work in progress. The struggle for a just and equal life for all is an on-going effort. We must never quit, never surrender, never give up.

Why Do We Possess Imagination?

When I was a child, my friends and I spent a lot of time playing "pretend. We played "army" often. Girls would play with dolls, pretending to care for them, as though they were real babies. Who hasn't pretended to be their favorite TV or film hero or heroine?

Imagination is said to be the ability to create. The ability to "lose" oneself in fantasy is strongest in children, but also exists in many adults. Our fascination with television, motion pictures, theater, and video games as a means of distracting ourselves from our present reality transcends culture and religion.

There appears to be a connection between our perception of an event as an image playing out before us in our sight and our sense of self. Television, motion pictures and theater duplicate this by showing us a "screen-image" of a false reality and with convincing acting, our brain is fooled into accepting what we are perceiving on stage, screen, and TV as real.

Our brains appear able to transcend the reality of the moment with an alternative reality—even if it is fake—as long as the image of the fake reality remains in our sensory perception. As soon as the false reality no longer exists, our brains revert back to our original state—the reality of the living room or theater audience.

Why?

We humans are also enchanted by anything that offers the opportunity to alter our sense of awareness of the reality of the present moment. Reality is relative and we humans are inexplicably drawn to whatever means are available, whether legal or not, to alter our sense of it. Whether by meditation, drugs, alcohol or other forms of sensory alteration, the human experience appears to be linked to an existence outside of our "normal" physical existence.

We seem to be "hard-wired" to imagination by our DNA. Has evolution adapted our consciousness to being able to transcend the reality of the moment? Or has the Universal Mind used evolution to apply the creative power of consciousness to transform our physical reality to that which we imagine in our minds?

Why?

The older we get, the more our ability to fantasize is lost. We lose the ability to fantasize and create. Is it due to hormonal changes of puberty? Does our minds normally lose their ability to create as we age amid the routine of day-to-day living in a reality that is reinforced to us with every breath we take and with every beat of our hearts?

Our hearts are the back-beat to every breath we take…

Breath, thump-thump, thump-thump…

Breath, thump-thump, thump-thump…

Together, each breath and heartbeat form the metronome of our existence in our physical world and the physical nature of our reality. We carry this metronome around inside us, every day, everywhere we go, until the day we die. Together, they form a constant reminder that our consciousness is held captive within a physical form, alien to its normal existence.

This metronome ties our consciousness to our bodies; centers and locks it in place within our bodies. But if we stop the heart, if we stop breathing, the metronome is no more. There is no longer anything to hold the consciousness within the physical brain. It is released.

Our mind's ability to use imagination and its instinctive urge to yield to altered states of awareness should be a clue as to our relationship with all that is nonphysical and not perceivable through our five senses.

Is There a Heaven or a Hell?

Simply put, there is no physical "heaven" or "hell." Each is a religious fantasy created by men who had no personal connection to Yeshua or his disciples. These were desperately ambitious, arrogant men at the Council of Nicaea who sought to transform a self-destructing Holy Roman Empire into the Holy Roman Catholic Church. Their sole purpose was to acquire the vast wealth and political power of the crumbling Roman Empire for themselves.

Posing as "religious authority" for the last two thousand years, these charlatans and their ideological descendants have manipulated our biological and psychological fear of death to give them our money and control our lives.

As long as we fear death, we give them and their evil empire fake-religious authority power over us. Deny them that fear and they and their power will vanish like the morning fog fades away in the morning Sun.

It is necessary that you understand the enormity of their deception. In ancient times, the most obvious method of seizing wealth and land was at the point of a sword. Those willing to kill for wealth and power fought each other for the right to rule.

Once the war was over, however, it became necessary to control the masses. Controlling the masses at the point of a sword is not recommended if one wishes to remain in power very long. The beneficent ruler ruled and lived longer than the ruthless dictator.

But even the most beneficent ruler refused to educate the masses to think for themselves. That would risk enlightening them that they could rule themselves—an unacceptable option if one wished to get rich off them and all rulers did so wish to get rich off their subjects.

To control the masses, rulers chose to indoctrinate their subjects in a religion which reinforced acceptance of the status quo and subservience to authority. The common masses were herded just like cattle. Religious indoctrination and acceptance of dictatorial authority go hand-in-hand.

There are always rebels and dissidents. Subservience to such evil authority carries no weight unless such internal dissent carries a punishment that deters others from rebelling. Executing radical rebels creates hatred and subversion and risks revolution by creating martyrs. Imprisonment does much the same.

The most effective means of controlling the masses is to brainwash them to be their own mental jail-keeper, so to speak. Whenever the indoctrinated masses encounter a radical idea that might threaten the ruling order, their minds automatically shut down on their own rather than risk going to hell for all eternity. That is, in effect, the current situation in America today. It is all part of the Christian myth.

However, within the Universal Mind, there is no 'heaven" or "hell." One's enlightenment is determined by the lessons we learn during each of our lives. We create our own "heaven" and "hell" by the choices we make during our lives. Good choices leac toward calm, drama-free lives where serenity is its own reward. Poor choices, on the othe hand, lead to a life of mental chaos and its unending whirlpool of self-inflicted suffering.

Either way, the choice is, inevitably, your own.

What is Karma ?

Karma is the conscious expression of Newton's Third Law of Motion: to every actio there is an equal and opposite reaction. Thought is the origin of all we create. Thought beget all human behavior. What we put out in our lives as thought and the emotions tha spring from such thought, return to us in the future.

This perspective equates thoughts with force. Thoughts are, therefore, forms of energ which must be in balance and under control for one to be positive and productive.

Thoughts compel action or nonaction. Therefore, thoughts are things. Thoughts matter.

Technology has changed the world, but it cannot change human behavior. Despite ou technological and scientific prowess, humanity remains trapped in the swirling seas of ou own, personal chaos. Too often we live out our lives as though we have been cas overboard from a sinking ship without knowing how to swim. We flounder, then sin beneath the waves of our own uncontrolled emotions.

Caught within a self-reinforcing cascade of calamity, we blindly ricochet from life t life to life. Then we suddenly awake one morning wondering why we are so bound to ou life's drama that we can never find peace. It should come as no surprise. But this cycl repeats itself because we lack insight into the uncontrolled nature of our minds.

One cannot evolve spiritually until one has consciously rejected this chaos as a life-style. To reject a thing requires one to first see it, to know it exists in one's life. Self-knowledge is difficult to attain, especially when one's thoughts are consumed by the chaos they are drowning in. But it is possible.

The world will never find peace until its people tire of their own self-made and self-reinforced chaos. Furthermore, the world will never be free from war until its people reject the violence within their own minds.

How can we do that? How can we find peace?

Considering Yeshua as a "personal savior" or as a mini-god, denies our own responsibility for our personal behavior and robs us of the just fruits of enlightened thought. He would not have wanted us to regard him in that way.

Yeshua taught us to save ourselves by our own enlightened thoughts and our own honorable deeds toward each other. We should recognize him, however, for his teaching our ancestors through simple analogies so unsophisticated shop-keepers, herdsmen, fishermen and soldiers of his time could understand how his path to personal peace and enlightenment.

Yeshua respected Karma enough that he made it a major theme of his Sermon on the Mount. Below are just a few lines from the very beginning of his sermon. As you read them, notice how every enlightened, positive thought results in a positive result and how every positive, enlightened action results in a positive reaction:

"Blessed are the poor in spirit; for theirs is the kingdom of heaven.

Blessed are they that mourn, for they shall be comforted.

Blessed are the meek, for they shall inherit the earth.

Blessed are they which do hunger and thirst after righteousness, for they shall be filled.

Blessed are the merciful, for they shall receive mercy.

Blessed are the pure of heart, for they shall see God.

Blessed are the peacemakers, for they shall be called the children of God.

Blessed are they which are persecuted for righteousness' sake, for theirs is the kingdom of heaven…"

Matthew 5 KJV

Credit: kingjamesbibleonline at kingjamesbibleonline.org

These are just a few examples of the many, subtle references to Karma Yeshua made during his Sermon on the Mount. The principle is as simple as it is pure: whatever one evokes in thought or deed, is returned in equal measure in this and all future lives. Positive thoughts create good deeds. Good deeds create good Karma. Good Karma creates an evolved future life here on Earth.

Spiritual evolution, like biological evolution, requires mutation. But where biological evolution requires a structural mutation of genetic code, spiritual evolution requires a structural alteration of one's own self-awareness. Such enlightenment requires an inner revolution of one's own perspective toward oneself.

This evolutionary enlightenment is hindered by negative Karma caused by a past-life (or lives) where negative emotions and a lack of self-knowledge controlled one's life experiences.

Many are trapped in lives drowning in the emotional chaos of their own making. Chaotic lives seek out other chaotic lives for validation and false solace. Such domestic relationships result in drama and, all too often, the tragedy of violence. Much of the violence and conflict in the world arise out of the undisciplined minds of those trapped in a seemingly endless cycle of personal drama.

The peaceful mind is like the straight rope. When such a straight rope is laid out, end to end, one can see its beginning and its ending. When the mind is as peaceful as this rope is straight, one receives clarity of thought, uncluttered by emotion. There is but one clear, straight line of conscious intention. Only the straight rope can see from one end to the other. This is the secret of enlightenment.

Negative thoughts, and the negative Karma such thoughts create, twist and turn the rope into knots. Such a knotted rope makes it impossible to see it from one end to the other. Enlightenment is not possible.

Sadly, this is the case for most of us. Our Guides create life gravity centers that force us to face these "knots" in our consciousness and through a balance of love and self-discovery, we grow in self-knowledge enough to unbind these knots. Only then can we evolve spiritually. Only then can we progress toward enlightenment. Only then can we hear in our mind's ear what the Universal Mind is speaking to us.

Why Does the Universal Mind Want Contact?

Earlier, in Chapter 20, we discussed how CERN is likely seeking direct contact with the Universal Mind. The question is whether the Universal Mind wants or cares enough about us to respond. Why should it?

Our world is facing a critical time in its history. As this book is being written, humanity stands at the precipice of chaos and catastrophe. Soul-crushing dictators in North Korea, China, Russia, and Iran, are scheming for war, slaughtering innocents, and imprisoning thousands of dissenters.

North Korea's Kim Jong-Un is launching missiles to bully South Korea, Japan and the United States while he continues to develop a nuclear weapons arsenal at the same time he starves his people to death. His master, China's Xi Jinping continues communist China's brutal genocide of peaceful Tibetans and Uighurs and repression of freedoms in both Hong Kong and mainland China. Russia's Putin is plotting a second invasion of Ukraine as he and his henchmen coldly assassinate his opponents. Iran's power-mad clerics continue to build their own nuclear weapons as they support any terrorist willing to attack Israel.

Meanwhile, the world continues to battle Covid-19 and climate change while looking the other way as impoverished people in Third World nations struggle just to survive from day to day.

And America? America is distracted by its own self-destruction, as the world churns and burns around it. America is no longer a beacon of hope, vision, and resolve to the world. Its social cohesion lies in tatters over its original sin of slavery and entrenched, socio-economic racism, police brutality and injustice. America's democracy is increasingly under attack by Republican insurrectionists, coup-plotters, white-supremacist militias, and propagandists.

Lastly, the world's two major religions, Christianity and Islam, have both lost whatever spiritual authority and moral influence either one possessed. Christianity's issues have been discussed extensively throughout this book.

Many of Islam's clerical leaders have made the fundamental error of linking their religion with anti-Semitism, human rights violations, enslavement of women and children, and the destruction of both the United States and Israel. There is nothing spiritually redeeming in any of these goals. Their goals are as spiritually illegitimate as they are irrational.

Yes, times are changing. The world is at increasing risk of a violent end. But there is hope. Even now, more and more individuals are becoming aware of their connection to the Universal Mind and its connection with all life.

This increasing awareness is reflected in the ever-growing resistance to the evils facing the world as well as the importance of conserving our natural resources. The battle for the soul of the world continues. Your help is urgently needed.

But before you commit yourself to joining the Universal Mind's force for good, you should know what the Universal Mind stands for. The following should give you an idea:

First, there is no separation between us and the Universal Mind. If you have accepted anything I have written in this book about the relationship between photons and the Universal Mind, then you will realize that once you strip away the transitory nature of the physical body, all that we are is what we have always been—pure consciousness as entangled photons.

The Universal Mind is not a religious figure or entity and should not be worshipped as such. It is us. We are it. Our five senses have imprisoned us with a false view of our relationship with the universe. Our senses lie to us. We are not separate. We are one with all that exists.

The Universal Mind plays by the rules. It must obey the same laws of physics that the rest of the universe must obey. While quantum entanglement and the relationship of nonlocality between all beings allows for what we term "miracles," the Universal Mind has limits as to what it can do.

The Universal Mind cannot prevent acts of free will, even if those acts of free will are negligent or malicious. The drunk driver who chooses to cross over into your lane will kill you regardless of what the Universal Mind wants. If you do not take your anti-Covid vaccine, you will risk death regardless of what the Universal Mind wants.

You have made your own choice. You have made your own Karma. You used your free will to choose the wrong path. Do not expect any miracles to save you from yourself.

What Are Miracles?

Miracles do exist. They do happen. But their occurrence is not the result of religious faith. Religion is not the reason people pray or meditate. These acts are the result of a personal, spiritual relationship between the individual and his or her "maker." This relationship exists independent of ANY religious involvement.

The science of miracles is simple to explain. Thoughts are things. Thoughts have energy. Nothing occurs or fails to occur in our lives which we did not think about first. Thought opens a channel of creative energy directly from the Universal Mind. The more focused and concentrated the connection between the individual and the Universal Mind, the more powerful the creative force which emanates from the Universal Mind.

A miracle is "merely" a thought transformed into physical reality. You don't need to be a saint or a prophet or Yeshua to receive a miracle. You don't need to throw your hard-earned money away with some fraud evangelist pretending he can turn water into wine for you—if you only give him a "donation."

Miracles, the real kind, are free. All they require is your ability to "touch" the Universal Mind with your thoughts and mental images of what you need. And that is the catch. You may pray for something you think you "need," when you really don't. There is a difference between what you "want" (what you "think" you need in the moment) and what you really need from an objective or spiritual point of view. The Universal Mind is not *"Alice's Restaurant."*

Credit: Arlo Guthrie, "Alice's Restaurant," Apple Music, 1967. The reference is to the beginning lyrics: "You can get anything you want at Alice's Restaurant…

Speaking of miracles, what are they? Let's distinguish the term, "miracle," from the act itself. The term "miracle" is offensive. To call something a "miracle" cynically plays upon the fears of the weak and the ignorant. They have been raised by the Church to associate "miracles" with the unseen "magic" of a fearsome god holding over us the threat of an eternal hell if we do not submit to its rules.

The Universal Mind has its own "rules"—natural laws. From these natural laws we derive the following principles:

Universal Mind Principles:

1. All life is based upon consciousness.
2. All consciousness is eternal.
3. All life is sacred and all self-aware life is to be protected.
4. Except in defense of self or another, all violence is wrong.
5. Thoughts are things. What you think toward others is returned to you. Treat others as you would have them treat you.
6. Heaven and hell are but states-of-mind.
7. God is love and enlightenment. God is not to be feared.
8. God does not punish. Karma is its own punishment.
9. God does not judge. Karma is its own judgment.
10. Any religion which claims to be superior to others is wrong.

If you can apply these principles, then the concept of "miracle" ceases to apply and the application of visualized thought enhanced by the power of the Universal Mind will take its place. You will begin to understand that what the fakers of religion call "miracles" are merely spontaneous events of quantum nonlocality—consciousness in union with the Universal Mind.

We discussed quantum nonlocality earlier in this book. These events are like lightning. They are causally created by fixed universal laws. Magic or supernatural power has nothing to do with it.

When we "pray" or meditate we seek a connection with the Universal Mind. True, based upon your upbringing and education, you may believe you are praying to a god or a human you revere. But those images are merely for your own benefit. These images are useful to give your mind a visual point of reference to focus upon, like visualizing a rose with your eyes closed during meditation.

What creates the "miracle" is your focused thought reaching the Universal Mind and creating an energy wave of quantum nonlocality to manifest your thought in your physical reality. That is what happened to me with my law school dream that became reality.

Let me explain it another way. Quantum nonlocality events (miracles) are like lightning strikes. They appear random or spontaneous, but in truth each lightning strike is the result of a specific set of atmospheric conditions.

"Most, if not all, lightning flashes produced by storms start inside the cloud. If a lightning flash is going to strike ground, a channel develops downward toward the surface. When it gets less than a hundred yards of the ground, objects like trees and bushes and buildings start sending up sparks to meet it. When one of the sparks connects the downward developing channel, a huge electric current surges rapidly down the channel to the object that produced the spark."

Source: "Severe Weather 101, Lightning Basics," National Severe Storms Laboratory, at nssl.noaa.gov. (No author, no date).

Like lightning, miracles occur due to the interaction of natural forces. Storm clouds produce positive and negative ions. Lightning results when these opposing charges overcome the insulating air separating them, releasing energy. They are not spontaneous but are the result of cause and effect.

The same is true for quantum events of nonlocality. The Universal Mind is source of infinite potential energy. Our meditation or prayer is a focused thought that attempts to create a channel with the Universal Mind. When such a connection is made, our consciousness becomes the "spark" that the Universal Mind connects with allowing the "miracle" to occur.

Therefore, miracles are a two-step process. First, there must be a connection between the mind of the person seeking the miracle and the Universal Mind. While our inherent creativity allows us to visualize a certain result, an energy force of sufficient power is still necessary to turn the visual image into a reality. This power comes from the Universal Mind itself and our minds must be clear and focused enough to serve as the lens through which this energy is directed.

In truth, then, there are no miracles, as such. There are simply events where a human mind has established a focused connection with the Universal Mind allowing the Universal Mind's energy to manifest whatever the person sought when they prayed or meditated.

These spontaneous events of quantum nonlocality should not be given any religious significance. They are not caused by any god. Such concepts only serve to block our conscious connection with what we have been part of since the Big Bang—the Universal Mind.

The Universal Mind and Human Rights

Consciousness is all and all is consciousness. Therefore, all living beings share the same origin—the same "maker." Therefore, all people are born with the inherent and inalienable right to be free and equal. So long as this freedom and equality does not intrude upon the rights of others, every person has the right to live their lives however they wish, regardless of gender, race, sexual orientation, political and religious belief, national origin or other demographic. Every human also has an inherent and inalienable right to free education and health care, safe housing, and sufficient food to eat.

It Is Up to You

Hopefully, this book has opened your mind to a new dimension of your inner self and to begin to apply its reality in your life. Your relationship with the Universal Mind should be one of loving compassion and joy toward all life. May all become fully aware of their connection with the Universal Mind—the "maker" of us all.

ILLUSTRATIONS

ILLUSTRATIONS CONTINUED

ABOUT THE AUTHOR

Michael A. Tewell received his master's degree in Kent State University's Graduate School and Honors College in 1979. He was admitted into graduate school after only his sophomore year. His master's thesis was *Ideological Terrorism: An Analysis of the Capability and Will of Transnational Terrorist Groups to Attack the United States*.

A Freemason like his grandfather before him, Tewell received his law degree in 1982 at the George Washington University School of Law in Washington, D.C. Tewell interned with both the United States Attorney's Office and the Enforcement Division of the United States Securities and Exchange Commission in Washington, D.C. while a law student.

A member of the Florida Bar since 1983, Tewell has been an attorney for over half his life, including one year as an Assistant State Attorney and twenty-five years defending the poor, mentally ill, juveniles and veterans with the Public Defender's Office. He has tried well over 100 jury trials including cases involving murders, DUI manslaughters, robberies, and other serious felonies and misdemeanors.

In addition, Tewell has represented clients at hundreds of nonjury trials and evidentiary motion hearings critical to the determination of legal issues. He has also won appeals before Florida's Second District Court of Appeals.

In 2004, Tewell was honored as the Assistant Public Defender of the Year in his jurisdiction. Before he retired, Tewell organized and taught a Trial Practice Clinic for new Assistant Public Defenders and enjoyed mentoring young lawyers.

Previous books by Tewell include ***Immortality: The Science of Forbidden Fruit*** and ***Reincarnation: The Scientific Theory***.

Born in Hamilton, Ohio, Tewell worked as an actor in NYC, a DJ for a West Virginia radio station, and as a newspaper reporter in both Texas and Ohio before attending Kent State University.

Retired from the Public Defender's office in 2016, Tewell began seriously studying classical and quantum mechanics, evolution, physiology, mathematics, and other sciences.

He dedicated himself to sharing with others what he experienced and learned during his life. He is the Executive Director and President of Universal Mind Foundation, Inc., a not-for-profit corporation created to educate the public about the Universal Mind and Reincarnation.

He now lives in Florida with his wife of 26 years (also a criminal defense trial lawyer) and their son. Tewell coached both club and high school soccer and his fondest sports memories were coaching his son and his son's teams.

Made in United States
Orlando, FL
04 November 2026

90208353R00187